Delivering Energy for Development

Praise for this book

'*Delivering Energy for Development* speaks directly to the "bottom up" energy access solutions that are needed in developing countries to get cleaner, more affordable and safer energy services to people. At the heart of this are diverse energy service delivery models that require a new generation of energy entrepreneurs, supportive government policies and an enabling framework including access to consumer financing to be successful. Practical Action has made a major contribution by correctly portraying people in these markets as consumers rather than beneficiaries.'

Kandeh Yumkella, Special Representative of the UN Secretary General
for the Sustainable Energy for All (SE4ALL) initiative

'This book gives an excellent overview of the main principles and success factors for overcoming energy poverty, and provides a framework for analysing and improving the delivery of energy services in developing countries. The book will help energy practitioners as well as policy makers to develop a better and more detailed understanding of the barriers that prevent universal access to modern energy services.'

Dr Carsten Hellpap, Program Director,
Energising Development (EnDev), GIZ

'*Delivering Energy for Development* provides timely and critical information on how to deliver energy to all, especially to the poorest global population. Energy being the enabler of all aspects of sustainable development, the book offers options to address the chronic global challenge of inequity in energy access in developing countries. It is a timely publication as nations embark on implementing the UN General Assembly declaration of 2014 to 2024 as the decade of Sustainable Energy for All.'

Stephen Gitonga, Policy Adviser on Energy,
United Nations Development Programme

'Ensuring access to modern energy for the poor is a central development challenge. This book clearly shows the benefits and limitations of the methods at work to deliver energy access, and points out where the organizations working in this space – governments, finance, entrepreneurs and civil society – can combine their strengths to greater effect.'

Jiwan Acharya, Senior Climate Change Specialist (Clean Energy),
Asian Development Bank

'This is another great addition to Practical Action's excellent resources on sustainable energy. Drawing on a host of experiences from across the globe, *Delivering Energy for Development* provides an excellent overview of current thinking on alternative delivery models and explores the key barriers to meeting the energy needs of the poor. Aimed at energy practitioners, this book will also be of strong interest to students and researchers working in energy and international fields.'

Ed Brown, Senior Lecturer in Human Geography, Loughborough University, and Co-Chair, UK Low Carbon Energy for Development Network

'This well-written book is an essential reference for all practitioners, policy makers and financiers active in the energy access space in developing countries. The book does an admirable job in presenting energy delivery models in a lucid manner supported by an excellent mix of case studies. Recommended.'

Dr Binu Parthan, Principal, Sustainable Energy Associates, Kottayam, India

Delivering Energy for Development
Models for achieving energy access for the world's poor

Raffaella Bellanca, Ewan Bloomfield and Kavita Rai

Practical
ACTION
PUBLISHING

Practical Action Publishing Ltd
25 Albert Street, Rugby, CV21 2SD, Warwickshire, UK
www.practicalactionpublishing.com

ISBN 978-1-85339-761-5 Hardback
ISBN 978-1-85339-762-2 Paperback
ISBN 978-1-78044-761-2 Library Ebook
ISBN 978-1-78044-762-9 Ebook

Book DOI: http://doi.org/10.3362/9781780447612

Bellanca, R., Bloomfield, E. and Rai, K. (2013) *Delivering Energy
for Development: Models for Achieving Energy Access for the World's Poor*, Rugby, UK:
Practical Action Publishing.

Since 1974, Practical Action Publishing has published and disseminated books and
information in support of international development work throughout the world.
Practical Action Publishing is a trading name of Practical Action Publishing Ltd
(Company Reg. No. 1159018), the wholly owned publishing company of Practical
Action. Practical Action Publishing trades only in support of its parent charity
objectives and any profits
are covenanted back to Practical Action (Charity Reg. No. 247257,
Group VAT Registration No. 880 9924 76).

Cover photo: Erecting a wind turbine in Patla Village, Phalamkhani, Nepal Photo
credit: Rakesh Shrestha, Practical Action
Typeset by Allzone Digital

The manufacturer's authorised representative in the EU for product safety is
Lightning Source France, 1 Av. Johannes Gutenberg, 78310 Maurepas, France.
compliance@lightningsource.fr

Contents

http://dx.doi.org/10.3362/9781780447612.000

Figures, tables and boxes

Figures

Tables

Boxes

About the authors

Raffaella Bellanca is Country Coordinator with International Lifeline Fund in Haiti, working on local production and commercialization of improved cookstoves. She has worked in the energy field for nearly 20 years, including experience as a cleantech entrepreneur, which inspired her interest in business models and their viability under different conditions. Her research has focused on combustion processes in power plants and car engines, and energy aspects in the social development sector. Her background is mainly in physics with a PhD from the University of Lund, Sweden, and she also holds a Master's degree in communication for development. She previously led the HEDON Household Energy Network.

Ewan Bloomfield is an International Energy Consultant with Practical Action Consulting, UK, with experience of a range of energy technology and policy issues in developing countries. He has more than 15 years of international development and engineering experience in Australia, Haiti, Central Asia, South Asia, Africa, and the Far East. He has an engineering background, including a PhD from KwaZulu-Natal University, South Africa, and currently specializes in household cooking and bioenergy, working in a wide range of areas including appropriate technology design and market development, from resource assessments and supply chain analysis to end-user awareness and policy development.

Kavita Rai is an energy specialist with 20 years of international experience in renewables, enterprise engagement and socio-economic development. She has been a researcher and an active practitioner promoting sustainable energy interventions, including working for GVEP International, IT Power, Camco and Practical Action. She is a trustee of the HEDON Household Energy Network and currently works for IRENA.

Annabel Yadoo has worked as an energy researcher and consultant for international NGOs Practical Action and Renewable World. She has a PhD in delivery models for decentralized rural electrification from the University of Cambridge and has conducted fieldwork in Peru, Nepal, Kenya, Nicaragua and Mozambique.

Acknowledgements

The idea for this book was born in 2010 out of the Decentralised Energy for Livelihoods, Environment and Resilience (DELiVER) group in London initiated by the International Institute for Environment and Development (IIED), Global Village Energy Partnership (GVEP) International, Practical Action, and HEDON Household Energy Network. The intention was to share the experiences of energy access delivery in a number of countries around the world, for further discussion and understanding of lessons learned. The book is an attempt to define and explain energy delivery for the poor, pulling together the acquired knowledge of the authors, as well as those of other experts, and to further the debate especially within the context of the UN's Sustainable Energy for All (SE4All) initiative.

This book has been the result of voluntary collaborations from the main authors, who believe that a much stronger analysis of the various elements of energy delivery models is required, and who hope that this book will challenge and/or inspire others to continue the task of researching and sharing their findings on energy access delivery with us and the wider energy practitioners' community.

Our warmest gratitude goes to Emma Wilson from IIED, who helped provide context and critical input into all areas of the book; Annabel Yadoo, who authored the original version of the introductory chapter; and Mattia Vianello from Practical Action. In addition, thanks go to Jonathon Rouse from HED Consult, Teodoro Sanchez from Practical Action, and Steven Hunt (formerly with Practical Action but now energy adviser for the UK's Department for International Development), who helped shape the original concepts of energy delivery models, and some of the case studies.

Thanks also go to the two reviewers: Andrew Barnett from The Policy Practice and Andrew Scott from the Overseas Development Institute (ODI), UK for their very in-depth and insightful guidance and feedback on the direction of the book.

Additional thanks goes to several other people who have provided valuable time and input to the book, including Martin Bounds, Mariana Gallo, Ben Garside, Essam Yassin Mohammed, Laya Taheri, Rachel Godfrey Woods, and Sarah Matthews.

Finally, thanks goes to the support provided by the Policy Innovation Systems for Clean Energy Security (PISCES) bioenergy research project which has been funded by DFID.

Acronyms and abbreviations

AC	alternating current
ACCI	African Clean Cooking Initiative
ADB	Asian Development Bank
AfD	Agence Française de Développement
AfDB	African Development Bank
AGECC	Advisory Group on Energy and Climate Change
AMADER	Agence Malienne pour le Développement de l'Energie Domestique et de l'Electrification Rurale
AMC	advanced market commitment
APCPDCL	Andhra Pradesh Central Power Distribution Company Limited
ARE	Alliance for Rural Electrification
ASCI	Administrative Staff College India
ASER	Agence Sénégalaise d'Electrification Rurale
AU	African Union
BAAC	Bank of Agriculture and Agricultural Cooperative
BHS	Bosch and Siemens Home Appliance Group
BKK	Badan Kredit Kecamatan
BMZ	German Federal Ministry for Economic Cooperation and Development
BoP	base of pyramid
BUD	ank Rakyat Unit Desa
CAFOD	Catholic aid agency for England and Wales
CBO	community-based organization
CDM	clean development mechanism
CEADS	Consejo Empresario Argentino para el Desarrollo Sostenible
CER	certified emissions reduction
CEPALCO	Cagayan Electric Power and Light Company
CFL	compact fluorescent lamp
CIDA	Canadian International Development Agency
CNFL	Compañía Nacional de Fuerza y Luz
CO$_2$	carbon dioxide
COMESA	Common Market for Eastern and Southern Africa
CSP	concentrated solar power
CSR	corporate social responsibility
CREE	Community Rural Electric Entities
DEEP	Developing Energy Enterprise Programme
DELiVER	Decentralised Energy for Livelihoods, Environment and Resilience
DfID	Department for International Development

DHC	district heating and cooling
DRC	Democratic Republic of Congo
DTR	digital transfer ratiometer
EAC	East African Community
EADB	East African Development Bank
EBRD	European Bank for Reconstruction and Development
ECA	export credit agency
ECC	energy contract company
ECOWAS	Economic Community of West African States
EnDev	Energising Development
ESCO	energy service company
ESMAP	Energy Sector Management Assistance Programme
EU	European Union
EUEI-PDF	EU Energy Initiative Partnership Dialogue Facility
FAO	Food and Agriculture Organization
FDI	foreign direct investment
FRES	Foundation Rural Energy Services
GDP	gross domestic product
GEF	Global Environment Facility
GIZ	German Agency for International Cooperation
GNESD	Global Network on Sustainable Development
GSI	Global Subsidies Initiative
GVEP	Global Village Energy Partnership
GW	gigawatt
GWEC	Global Wind Energy Council
HH	household
HPS	Husk Power Systems
IADB	Inter-American Development Bank
IAP	indoor air pollution
ICE	Instituto Costarricense de Electricidad
ICT	information communication technologies
IDA	International Development Association
IDB	Islamic Development Bank
IDE	International Development Enterprises
IEA	International Energy Agency
IFC	International Finance Corporation
IIED	International Institute for Environment and Development
IMF	International Monetary Fund
IPSE	Instituto de Planificación y Promoción de Soluciones Energéticas para las Zonas no Interconectadas
IRENA	International Renewable Energy Agency
ISPRE	International Science Panel on Renewable Energies
IT	information technology
KEFRI	Kenya Forestry Research Institute
KFS	Kenya Forestry Service

kg	kilogram
KJL	Kampala Jellitone Suppliers
kV	kilovolt
kW	kilowatt
kWh	kilowatt hour
LED	light-emitting diode
LPG	liquid petroleum gas
LYDEC	Lyonnaise des Eaux de Casablanca
m³	cubic metre
MDG	Millennium Development Goal
MEGA	Mulanje Electricity Generation Authority
MFI	microfinance institution
MIGA	Multilateral Investment Guarantee Agency
M-Pesa	mobile money
MSME	micro-, small- and medium-scale enterprises
MW	megawatt
M&E	monitoring and evaluation
NACEUN	National Association of Community Electricity Users-Nepal
NASA	National Aeronautics and Space Administration
NEA	Nepal Electricity Authority
NEPAD	New Economic Partnership for Africa's Development
NGO	non-governmental organization
OBA	output-based aid
ODA	overseas development aid
ODI	Overseas Development Institute
OECD	Organisation for Economic Cooperation and Development
PBS	Palli Bidyut Samity
PEP	private energy provider
PERMER	Project for Renewable Energy in Rural Markets
PISCES	Policy Innovation Solutions for Clean Energy Solutions
PPA	power purchase agreements
PPEO	*Poor people's energy outlook*
PPO	pure plant oil
PPP	public–private partnerships
PRSP	Poverty Reduction Strategy Paper
PV	photovoltaic
RBF	results-based finance
REA	Rural Energy Agency
REB	Rural Electrification Board
R&D	research and development
RECS	Rural Electrification Collective Scheme
REDD+	UN collaborative initiative on Reducing Emissions from Deforestation and Forest Degradation in Developing Countries
RERED	Renewable Energy for Rural Economic Development
RESCO	Renewable Energy Service Company

ROSCAs	rotating savings and credit associations
RRUSPL	Rural Renewable Urja Solutions Private Ltd
SACCO	savings and credit cooperative
SADC	Southern African Development Community
S³IDF	Small Scale Sustainable Development Fund
SEf	Solar Energy foundation
SE4All	Sustainable Energy for All
SEM	small enterprise management
SEMA	Sustainable Energy Markets Acceleration project
SIDA	Swedish Institute for Development Cooperation
SME	small and medium-sized enterprise
SNV	Netherlands Development Organisation
SPP	small power producer
STEG	Société Tunisienne de l'Electricité et de Gaz
SWH	solar water heating
TEDAP	Tanzania Energy Development and Access Project
UAE	United Arab Emirates
UN	United Nations
UNCDF	UN's capital investment agency
UNDP	United Nations Development Programme
UNEP	United Nations Environmental Programme
UNFCCC	United Nations Framework Convention on Climate Change
UN-GEF	United Nations Global Environment Facility
UPPF	Unified Petroleum Price Fund
USAID	United States Agency for International Development
VSPP	very small power producer
WBG	World Bank Group
YYAG	Youth to Youth Action Group

CHAPTER 1

Introduction: The energy access and delivery challenge

Annabel Yadoo, Raffaella Bellanca, Ewan Bloomfield and Kavita Rai

Access to clean energy is projected to remain an issue for poor people, particularly those living in rural areas, for years to come. In recognition of the importance of energy access for human development, the UN-led SE4All initiative has set a target to achieve universal energy access by 2030. This book aims to give energy practitioners guidance on how energy delivery models need to change in order to fully overcome energy poverty, including important lessons from well-known case studies. It also provides in-depth analysis of the barriers to sustainable, affordable and effective energy access and delivery, particularly for the poor, with recommendations for practitioners on overcoming these barriers.

Keywords: development, energy delivery model, energy poverty, modern energy services

Billions of people in the world remain trapped in energy poverty, through being unable to obtain the quantities of energy they require, in clean and usable forms. Energy practitioners continue to be faced with the challenge of designing systems to deliver access to a range of modern energy services, particularly where poverty and lack of infrastructure makes it most difficult. More recently, issues relating to climate change, and the notion of sustainable and secure supplies of energy for a still growing world population, are also adding complexity to the challenge. Worldwide, energy poverty remains staggeringly high; in 2011 over 1.3 billion people – 19 per cent of the world's population – lacked access to electricity, and 2.7 billion people – 39 per cent – still relied on traditional three-stone fires for cooking (OECD/IEA, 2011). The vast majority of these people (over 95 per cent) live in rural areas of sub-Saharan Africa and South Asia; worldwide 84 per cent of people who lack access to electricity in their homes live in rural areas (IEA, 2011). Based on current policies and future demographics, the International Energy Agency projects that over 1 billion people will still lack access to electricity in 2030, of which 85 per cent will live in rural areas, mostly in sub-Saharan Africa, India and other parts of developing Asia. Similarly, the

http://dx.doi.org/10.3362/9781780447612.001

number of people relying on the traditional use of biomass for cooking is expected to remain at 2.7 billion in 2030, of which 82 per cent will live in rural areas (IEA, 2011).

There is a growing body of evidence which demonstrates that energy is a vital catalyst for human development, notably through improving health, education, food security, gender equality and the ability to earn a living. The high incidence of energy poverty thus provides a growing impetus for energy practitioners to increase their efforts on achieving the longed-for goal of clean energy access for all (DFID, 2002; GNESD, 2007). Since 2010, IEA has dedicated a chapter of its annual *World Energy Outlook* to the topic of energy poverty and how to achieve universal access to energy. In recognition of the importance of energy access for economic and human development, in September 2010 the United Nations (UN) launched an ambitious goal to achieve universal energy access by 2030, which has now evolved into the Sustainable Energy for All initiative (SE4All, n.d.). The year 2012 was also declared the UN International Year of Sustainable Energy for All, to try to help kick-start the initiative and push energy poverty up the global agenda.

This book has been written primarily for energy practitioners, and aims to provide a more integrated analysis of how energy has been delivered, and how this delivery needs to change in the future to fully overcome energy poverty. It gives an overview of what is meant by energy delivery, including a new definition of the term 'energy delivery model', as well as presenting important lessons from a number of well-known case studies of existing energy delivery models. It also provides guidance for practitioners on the key aspects of delivery model design. These aspects are integral to the success or failure of energy programmes, projects, and enterprises seeking to help people bring themselves out of poverty through access to energy in middle- and low-income countries. Finally, it provides in-depth analysis of some of the key barriers to sustainable, affordable and effective energy access and delivery, particularly for the poor, with recommendations for practitioners on overcoming these barriers.

Modern energy services for poverty reduction

Energy services are intrinsic to our basic survival. The ability to harness energy to meet human needs is not new; for hundreds of thousands of years, people have been burning wood to generate heat, light and warmth, and using the power of animals for transportation and harvesting food. Human societies have progressively learned to power their activities with increasing amounts of energy. This energy is then used directly for work (e.g. activities such as transportation of water or materials) or is converted from its natural source – wind, water, the sun, hydrocarbons and minerals – into mechanical power, transport, electricity, heating and so forth, as outlined in Box 1.1.

Box 1.1 Energy on Earth

The sun (a burning ball of hydrogen and helium) is the most abundant source of energy in our solar system. Solar energy travels to earth and is captured by green-leafed plants, which store the energy in their woody trunks, branches, roots, seeds and leaves – with fossil fuels being their historically concentrated remnants. Humans have been using huge quantities of a range of fossil fuels, ranging from solid fuels (coal and peat), to liquids (diesel, petroleum, and paraffin), to natural gas, to power their development; and still continue to do so.

The sun's energy can also be captured directly, as thermal or photovoltaic energy. The sun also heats and cools the earth's landmasses and seas, and, as the solar system's largest mass, it also significantly affects the planets' gravitational movements. The combination of these two activities produces most of the earth's other energy sources, including wind, wave and tidal energy. Hydropower originates in the gravitational flow of precipitation that falls at high altitudes as rain or snow and flows towards the sea. Geothermal energy comes from thermal energy produced by the earth's core which rises to the crust and can be captured.

More technological ways of capturing energy to optimize work have been developed with time, from wind mills and sailing boats, to the industrial revolution, fuelled by the discovery of coal and petroleum. The convenience, and greatly increased productivity, of mechanized agricultural practices, and other modern energy services, have changed the way many societies are able to live, and have gradually been expanded to new populations and geographies around the world. Today, most people desire a reliable supply of electricity to mechanically process their agricultural products, to listen to the radio, to recharge their mobile phones, to watch TV, and to power a whole range of other electrical appliances. They also require energy to power a range of modes of transportation, a better quality of light to see by, particularly in the evenings, and efficient ways of cooking which do not emit unhealthy emissions, nor require hours spent gathering firewood. Many poor people in the world do not have access to these energy services, and this is finally being recognized as an injustice which energy practitioners need to overcome. It is believed that overcoming this injustice will require new and innovative energy delivery models aimed at the poor, which will be analysed in detail in this book.

Although people require access to energy, it cannot just be in any form. Households and small businesses require energy in modern forms that allow them to carry out the activities they need and desire to, in convenient, healthy and sustainable ways, usually referred to as modern energy services. Electricity can provide the equivalent light-hours 70–160 times more cheaply than more traditional fuels such as candles or kerosene, once the quality of the light service (measured in lumens – the total quantity of visible light emitted by a source) has been taken into account (Foster et al., 2000). Liquefied petroleum gas (LPG) or efficient wood-burning stoves can displace inefficient three-stone fires (the traditional and cheapest cooking method used in most developing countries), which require long hours spent on fuel collection and contribute to deforestation and damaged health. According to the OECD/IEA (2010)

1.4 m people – mostly women and children – die each year as a result of inhaling smoke from traditional cooking stoves, approximately double the number of worldwide deaths from malaria.

As the UN Secretary-General Ban Ki-moon noted, access to modern energy services is the 'foundation for all the Millennium Development Goals' (MDGs) (UN, n.d.). In this context, modern energy services aim to provide a higher quality service in terms of light, heat, speed, etc., including being healthier (reduced harmful emissions and improved quality of light), reducing household expenditure, and increasing resource efficiency, to allow target populations to move onto sustainable technological pathways for their own development (Bazilian et al., 2010).

Table 1.1 details the contribution of modern energy services to the MDGs, particularly those related to healthcare, education, food security, poverty reduction, gender equality, and environmental sustainability. Work is currently being carried out to develop a number of post–MDGs development goals, and access to a range of modern energy services will continue to be an essential prerequisite for international development. In addition, analysis has shown that the greatest incremental benefit of modern energy services is received by those who are currently at the lowest levels of human development, and currently have the lowest levels of energy access (OECD/IEA, 2004). These people should therefore be the primary target of future energy delivery models which aim to promote development through increased energy access. It is these delivery models that this book aims to assess in detail, through a range of relevant case studies from around the world, detailed in Chapters 3 to 5.

Defining access to modern energy services

While modern energy services are increasingly recognized as a prerequisite for human and economic development (Munasinghe, 1987; Perlack et al., 1990; DFID, 2002; IEA, 2004; GNESD, 2007), there is still no universally accepted definition as to what is actually entailed by access to modern energy services, while the UN's definition of universality also creates some confusion, without clarity on whether all of a household's energy needs (cooking, electrification and mechanical power) must be met in order for the goal to be achieved. In future it is hoped that the research currently being conducted by the World Bank Energy Sector Management Assistance Programme (ESMAP) will help develop a global tracking framework for energy access. The UN is also planning to launch specific targets for universal energy access which should provide greater clarity.

Practitioners use varying assumptions for the 'definition of basic needs, fuels used and the efficiencies of energy-using technologies', preventing like-for-like comparisons of the current status and trends of energy poverty in different countries (Practical Action, 2010). Moreover, several definitions focus exclusively on the household level, excluding access for businesses and community services (for example, schools, health and community centres) yet these also have a significant role in reducing poverty. It should be noted

Table 1.1 Potential impacts of modern energy services and the Millennium Development Goals

Water pumping	Water disinfection	Cooking and water heating	Crop drying/cooling	Lighting and powering electrical equipment	Ambience heating/cooling	Transportation
• Clean water for drinking and washing improves hygiene (MDGs 4, 5) • Women and children spend less time fetching water, more can attend school (MDGs 2, 3) • Irrigation can increase the volumes and diversity of produce, improving incomes and local diets (MDGs 1, 4, 5) • Cultivation of produce during the dry season can fetch higher prices if sold, and save households money if consumed (MDG 1)	• Health improves as gastro-intestinal infections are reduced (MDGs 4, 5)	• Modern forms of cooking (e.g. gas and efficient biomass stoves) can reduce the amount of smoke inhaled and improve the health of women and children (MDGs 4, 5) • Women spend less time and travel less distance to fetch fuel, personal safety can improve (MDGs 3, 5, 6) • Reduced de-forestation and associated degradation (soil erosion, landslides, desertification, decreased biodiversity and fewer carbon sinks) (MDG 7)	• Preserving produce through the non-harvest season helps farmers feed their families throughout the year and sell produce at times when they can fetch a higher price (MDG 1)	• Health improves due to the higher retention of professional staff, refrigeration of vaccines, improved lighting, and the use of other electrical equipment (MDGs 4, 5, 6) • Education improves due to the higher retention of professional staff, improved lighting for home study, and computers and other learning aids (MDGs 2, 4) • Improved communications with potential impact on income (MDGs 1, 2, 8) • New or improved income-generation possibilities, local production of 'value-added' goods, and small-scale industry (MDG 1) • Greater comfort, ease and entertainment in the home, security from street lighting (MDG 3)	• Greater comfort (MDG 1)	• Improved communications (MDGs 1, 2, 8) • Improved healthcare, education and income-generation possibilities due to easier access to neighbouring communities and markets (MDGs 1, 4, 5) • Increased ability for healthcare and education professionals to reach the community (MDGs 2, 4, 5, 6)

Note: MDG 1 – End poverty and hunger; MDG 2 – Universal education; MDG 3 – Gender equality; MDG 4 – Child health; MDG 5 – Maternal health; MDG 6 – Combat HIV/AIDS; MDG 7 – Environmental sustainability; MDG 8 – Global Partnership. All energy services can contribute to MDG 7 if low-carbon sources are used sustainably (Yadoo et al., 2011)

that this focus is beginning to change through recent publications such as the United Nations Development Programme's 2012 report on energy and employment (UNDP, 2012) and Practical Action's *Poor people's energy outlook 2012* (Practical Action, 2012), which focused on energy for earning a living. Box 1.2 outlines what it understood by the term 'access to modern energy services' in the context of this book.

Box 1.2 Modern energy services

The IEA defines access to modern energy services as 'a household having reliable and affordable access to clean cooking facilities, a first connection to electricity [it should be noted that this electricity can come from either a grid, mini-grid or off-grid electricity generating system, rather than the more traditional assumption that electricity only comes through a connection to the grid], and then an increasing level of electricity consumption over time to reach the regional average' (OECD/IEA, 2011, p12).

Initial electricity consumption levels (based on the types of appliances – fans, mobile telephones, light bulbs, and so on – that could be operated within its limit) are 250 kilowatt hours (kWh)/year for rural households and 500 kWh/year for urban households, and consumption levels are expected to reach a regional average within five years. By 2030 consumption is expected to be 800 kWh/year/capita) (OECD/IEA, 2011).

Clean cooking facilities are referred to as 'biogas systems, liquefied petroleum gas (LPG) stoves and advanced biomass cook stoves that have considerably lower emissions and higher efficiencies than traditional three-stone fires for cooking' (OECD/IEA, 2011).

Use of this term is important for emphasizing that people need access not to any type of energy, but to forms of energy that are clean and which allow them to use the energy productively to improve the quality of their lives. In addition, referring to the *use* of modern energy services emphasizes the importance of end-use technologies and the way that these technologies are employed to improve people's lives and livelihoods. End-use technologies include the appliances that all households require, such as clean burning cooking stoves, electric light bulbs, and efficient agricultural processing machines. As well as the energy itself, all people require access to a range of energy technologies, which will be detailed in Chapter 2. Access to these energy technologies allows households, businesses, and communities to use energy in the ways they desire to improve their lives in a number of ways.

They may also require support or training in making effective use of the technologies, and it is only when these aspects of access are addressed that the true impact of energy access can be achieved, and energy poverty can start to be effectively overcome. Organizations such as Practical Action have started to refer to this type of energy technology inequality as 'technology justice'.

To try to take this understanding further, the *Poor people's energy outlook* (PPEO) 2012 proposed the concept of 'Total Energy Access', which outlines a full range of energy services believed to be required by each household, as well as nine minimum standards for these energy services, as outlined in Table 1.2. These standards diverge from the IEA's approach by framing energy in terms of the energy services provided, rather than access to an electricity connection,

Table 1.2 The *PPEO* definition of Total Energy Access

Energy service		Minimum standard
1	Lighting	300 lumens at household level
2	Cooking and water heating	1 kg wood fuel or 0.3 kg charcoal or 0.04 kg of LPG or 0.2 litres of kerosene or ethanol per person per day, taking less than 30 minutes per household per day to obtain Minimum efficiency of improved wood and charcoal stoves to be 40% greater than a three-stone fire in terms of fuel use Annual mean concentrations of particulate matter (PM2.5) < 10 µg/m³ in households, with interim goals of 15 microgrammes (µg)/m³, 25 µg/m³ and 35 µg/m³
3	Space heating	Minimum daytime indoor air temperature of 12°C
4	Cooling	Food processors, retailers, and householders have facilities to extend life of perishable products by a minimum of 50% over that allowed by ambient storage All health facilities have refrigeration adequate for the blood, vaccine, and medicinal needs of local populations Maximum indoor air temperature of 30°C
5	Information and communications	People can communicate electronic information beyond the locality in which they live People can access electronic media relevant to their lives and livelihoods
6	Earning a living	Access to energy is sufficient for the start-up of any enterprise The proportion of operating costs for energy consumption in energy-efficient enterprises is financially sustainable

Source: Practical Action, 2012: 42

for example. They also do not specify minimum electricity consumption levels, which are too general and do not provide information about what is specifically required by end users (i.e. a range of energy service levels). If an assessment is carried out and a household meets all of the minimum energy standards, they are then considered to have total access to modern energy services, defined as Total Energy Access.

It should be noted that this list does not include access to mechanical power, including static and mobile shaft power (static shaft power refers commonly to mechanized pumps, mills, hand tools, etc., and mobile shaft power to vehicles). This type of power is difficult to relate to individual household usage as it is often more closely associated with community energy usage. This matrix of energy services and minimum energy standards is currently being reviewed by the international community, spearheaded by the World Bank's ESMAP, and is likely to be updated in the near future so as to be able to provide a universal measure of access to modern energy services.

Delivering energy for development

A wide range of experts and organizations, among them multilateral institutions, non-governmental organizations (NGOs), private companies and social enterprises, have been involved in designing, developing, financing, and implementing access to energy programmes, projects, and enterprises in the developing world since the 1970s. (Note that in this book, a 'social enterprise' is defined as an enterprise that uses commercial approaches to achieve sustainability of supply, but having social as well as financial goals.)

Box 1.3 provides a very simplified and brief history of how energy access programmes aimed at reducing energy poverty have gradually changed over time, particularly concerning the gradual shift towards more decentralized governance, privatization, and focus on rural areas.

Box 1.3 A brief history of access to energy programmes

Energy access programmes have taken a variety of forms over the years. During the early years (in the 1970s and 80s), many programmes tended to be state-led and top-down; governments centrally designed, implemented, and ran the energy systems. However, in order to better encapsulate the needs and preferences of the intended beneficiaries, access to energy programmes – especially after the 1980s – have gradually become more bottom-up and inclusive, incorporating user consultations and participation, and involving regional, provincial, and local bodies to coordinate, plan, and deliver the energy services. Rather than being involved in the programme design and implementation stage, under such approaches governments will often be expected to establish policies, rules and regulations, and to provide funding. A by-product of this shift towards greater inclusivity and responsiveness to local needs was the growth in the dissemination of decentralized, off-grid energy systems. Whereas centralized energy options such as grid extensions had been the norm, increasingly the importance of decentralized energy systems (such as pico- or micro-hydro plants, solar lanterns, solar home systems, diesel generators, or community wind turbines) were being recognized as a way of improving access in remote, rural areas.

There has also been considerable shifting back and forth between public and private-sector approaches, most notably during, and in the aftermath of, the widespread privatization and liberalization of energy sectors since the early 1980s. On the whole, the privatization of state utilities failed to increase access for households in rural areas (Cherni and Preston, 2007); in fact, tariffs in developing countries largely increased as a result of the reforms, rural electrification rates experienced sharp drops and electricity consumption declined (Haanyika, 2006). Although rural energy provision is not yet a sustainable business proposition in many developing countries, there is increased appreciation of the need to form public–private partnerships to increase energy access (see for example Boiling Point, 2010). Among other things, this can include the provision of output-based subsidies or start-up grants for private enterprises by public-sector institutions (governments, NGOs, and multilateral bodies) and the support of socially motivated equity investors or debt financing institutions.

Below is a very simplified summary of energy access programmes over the years:

- 1970s and 80s – Top-down, government-led approaches to energy programmes
- 1980s and 90s – Increasing privatization and liberalization of national energy sectors
- From 2000s – Decreasing rate of privatization, awareness of need to adopt a targeted approach to rural electrification. Many developing countries established dedicated Rural Energy Agencies and Rural Electrification Agencies
- Present day – Increasing amount of public–private partnerships and the use of public subsidies to incentivize private-sector investment in rural energy access

Source: Reiche et al., 2000; authors' own research

As discussed in the following chapters, the technologies used to deliver energy vary as widely as the approaches adopted by the institutions – both private and public. Some adopt a sector-wide approach (for example, the Netherlands Development Organisation, SNV, which has focused on building up market sectors for biogas in developing countries around the world). Others have engaged in activities on a more local scale, either within low-income commercial markets (for instance, the UK-based NGO SolarAid which has established micro-franchises for off-the-shelf solar products through its Sunny Money programme in East Africa), or in areas that currently lie outside the potential range for market-based approaches (for example, Soluciones Prácticas, Practical Action's office in Peru, which has set up, and helps maintain, renewable energy mini-grids in remote off-grid areas which are not currently viable for private-sector investment).

The entire process of delivering energy services and products is often referred to as an energy delivery model. The model encompasses the entire chain, from the initial energy resource, either fossil fuel-derived or renewable, as well as the technology choices, implementation process, and surrounding support and governance infrastructure, including any long-term maintenance arrangements. The concept of the energy delivery model will be explored in more detail in Chapter 2.

In its simplest form, an energy delivery model contains the three energy elements: the initial energy source, the conversion equipment, and the final energy appliances the end users require, as outlined in Box 1.4.

Box 1.4 The three energy elements

The diagram here describes a simplified complete energy system, with the three main energy elements. From the original energy source, to the energy conversion equipment, through to the usable energy: the energy appliances which deliver the energy service to the end user, the household, community institution, or business.

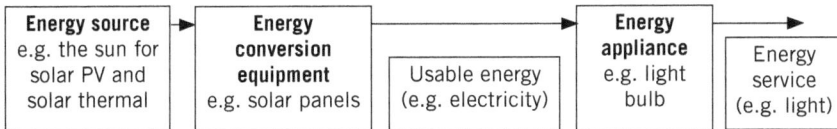

Energy source e.g. the sun for solar PV and solar thermal	Energy conversion equipment e.g. solar panels	Usable energy (e.g. electricity)	Energy appliance e.g. light bulb	Energy service (e.g. light)

Energy source: This is the original form in which energy is available, and includes wind, sunlight, water flow, wood, oil, etc. Sometimes the energy source can be used immediately, such as wood burned directly (or as charcoal) in stoves, but more often it needs to be processed using conversion equipment. Energy sources are either renewable or non-renewable.

Energy conversion equipment: This refers to the physical equipment used to convert the energy source into a useable form, such as electricity (e.g. turbines and generators) and mechanical power (e.g. wind mills). Following conversion, the energy can be stored for later use, such as in the form of batteries or processed biomass, such as briquettes.

Energy appliances: This is the equipment that is often required by the user in order to directly benefit from the energy service, and includes cook stoves, light bulbs, refrigerators, cars, etc.

These three energy elements are outlined in more detail in Table 1.3. The table shows both non-renewable and renewable energy sources; the range of conversion equipment, ranging from solar through to biomass and geothermal technologies; and the final end-user energy appliances for use with electricity, packaged fuels, and mechanical power.

Table 1.3 Energy sources, conversion equipment, and appliances

Energy source	
Non-renewable	e.g. coal, crude oil, natural gas, uranium (for nuclear power)
Renewable	e.g. the sun's rays and heat, woody biomass, biofuels, bioresidues, wind, waves, tides, running water, geothermal heat
Energy conversion equipment	
Solar technologies	e.g. photovoltaic (PV) solar home systems and solar lanterns; concentrated solar power (solar arrays); and solar thermal technologies (thermal water heaters, solar cookers, and driers)
Biomass technologies	biogas digesters, biomass gasification units, biomass burning power plants, and kilns
Liquid fuel technologies	power plants, engines, and generators
Mechanical energy technologies	wind, hydro and hybrid (combination) systems, ocean power systems
Geothermal technology	geothermal power stations, heating plants
Energy storage technologies	electrical energy in batteries
Energy appliances	
Electricity	light bulbs, electric cookers, sewing machines, refrigerators, televisions, mobile telephones
Gas, liquid and solid fuels	gas stoves and mantle lamps (for use with liquid petroleum, biogas, and piped gas), pressurized and wick kerosene stoves, wick lamps, ethanol stoves, traditional stoves, and improved biomass stoves (such as those fed with charcoal, dung, agricultural residues, briquettes or wood)
Mechanical power	mechanical mills and vehicles (tractors, cars, motorbikes), water pumps, treadle pumps, gravity ropeways, etc.

Source: Practical Action, 2012

The design and implementation of an energy delivery model determines the extent to which it achieves sustainable welfare benefits and reduces poverty for its target beneficiaries. All too often, energy access initiatives have increased access in middle-income markets, but inappropriate design or insurmountable barriers, often relating to cost and perceived value, have prevented them from reaching the very poor. For example the fairly recent and extensive dissemination of solar home systems in a number of developing countries including Bangladesh, India, and Sri Lanka, have mainly only been accessed by the middle class or less poor (Nygaard, 2009; Jacobson, 2007). Other initiatives have failed completely – for example, many thousands of solar cookers have been disseminated but are not used on a daily basis as they are not convenient and do not meet the needs of their intended users; solar cookers need to be placed directly in the sun, and the cooks generally do not find it comfortable to cook in such conditions. In addition, solar cookers are generally only able to produce the first cooked food by the early afternoon rather than the early morning, and cannot be used in the evening when households usually require them (Ahmad, 2001).

Despite efforts by governments to extend national grid access, millions of people worldwide living in grid-connected towns or villages, are still unable to receive electricity in their homes due to the prohibitively high cost of connection, or the lack of generation capacity and poor transmission infrastructure (Krishnaswamy, 2010). Therefore, despite over four decades of concerted effort on the part of governments, local and international organizations, and donors, in many developing countries only limited progress has been made in effectively delivering energy to the poor. The challenge to increase their energy access remains as great as ever.

Adding value through productive uses

All too often, the energy services or products themselves – for example, the provision of a grid connection, or the installation of a solar home systems or improved cooking stove – are considered to be the end goal of an energy access intervention. The need to look beyond lighting and cooking, and towards productive usage, has often been overlooked by practitioners. Many mistakenly assume that such development impacts will occur organically without the provision of additional seed capital, capacity building, and technology transfer. The authors contest this notion, and argue that access to energy should be seen as the beginning of an intervention process aimed at stimulating a wider range of impacts through the development of productive uses of energy services, and the application of energy within a broad range of welfare-improving services. This is where the real developmental benefits of energy access interventions lie.

For example, rather than stopping at the point of commissioning an electricity mini-grid, time and money should be invested in developing appropriate productive and socially orientated uses of electricity. These could include refrigerators for vaccines or dairy produce, carpentry workshops, agro-processing centres, computer centres at schools incorporating adult evening classes, and so forth. Similarly, by building up supply chains for the sale of fertilizers or purchasing electrical generators following the installation of biogas digesters, additional value can be locally generated and distributed among a wider range of stakeholders. It is important to note that these technologies need to be designed and installed in association with the availability of local resources, market linkages, and appropriate social processes as required.

Energy experts have been working to highlight the importance of ensuring that livelihood development becomes the focal point of market-driven energy delivery programmes, ensuring that energy services sustainably reach the marginalized but also promote socio-economic development. It has been stated that it is 'easier to make the profitable social, than the social profitable' (Khennas and Barnett, 2000: 18). To assist in this process, universally accepted guidelines need to be developed to measure the direct and indirect impacts of energy interventions. These measurements can then help with the setting of targets to ensure programmes are effectively designed and delivered, through careful planning, monitoring, and improvement. Work has already begun in this area, for example, the World Bank's 2008 Independent Evaluation Group; the ESMAP-led development of appropriate energy access indicators (supported by leading institutions); the SE4All initiative; the Global Alliance for Clean Cookstoves; and the World Bank's Africa Clean Cooking Initiative (ACCI), among others. As the learning from these initiatives is gradually embraced by all energy access practitioners, future energy delivery models can become more integrated with development goals. This should then lead to genuine reductions in energy poverty, increased livelihood benefits, and measurable long-term impacts on poverty.

Energy access and climate change

Anthropogenic climate change is a pressing global concern that is predicted to most severely affect the developing world where communities are least resilient (Zerriffi and Wilson, 2010). Its close association with recent energy production and use makes it impossible not to discuss climate change within the context of this book. The political and public awareness of climate change and its potentially devastating effects on many developing countries has significantly increased in recent years, and the move towards low-carbon energy production has risen up the political and donor agenda.

However, far from dramatically aggravating the climate problem, achieving universal access to modern energy services (as stated in the 'Defining access to modern energy services' section of this chapter) by 2030 would only result in an approximately 1 per cent increase in global energy consumption (Practical Action, 2010). This translates into an estimated 1.6 per cent increase in global greenhouse gas emissions, even if this was based only on on-grid electricity from fossil-fuel sources (Sanchez, 2010). In reality, as the UN Secretary General's Advisory Group on Energy and Climate Change (AGECC) points out, even this slight increase could be reduced in developing countries through energy efficiency measures and the use of renewable energy (AGECC, 2010). In addition, and particularly with regard to improved energy services for cooking, access to modern energy services is likely to have a positive overall environmental impact. For example, reducing the amount of deforestation (the depletion of carbon sinks); averting associated degradation such as soil erosion, landslides, desertification, decreased biodiversity, and the loss of ecosystems; and reducing black carbon, a potentially significant global warming trigger.

Access to modern energy services and the diversification of traditional energy options has the potential to significantly improve the adaptive capacity of poor communities, particularly those that rely heavily on their local environment (specifically agricultural and forest systems) for their livelihoods and welfare needs, as they are the ones likely to suffer the most from a changing climate (Johnson and Lambe, 2009). Developing countries have contributed little to global warming, as highlighted by Figure 1.1 which shows that the US, Europe, and China together produced approximately 55 per cent of all carbon dioxide (CO_2) emissions in 2008. Therefore developing countries should not be expected to bear the additional cost of climate change mitigation, including the opportunity cost of forgoing higher carbon technologies if they would better meet their development needs (Zerriffi and Wilson, 2010). Thus, while there is a desperate need to reduce greenhouse gas emissions, there is a growing consensus that low-carbon energy systems should only be prioritized

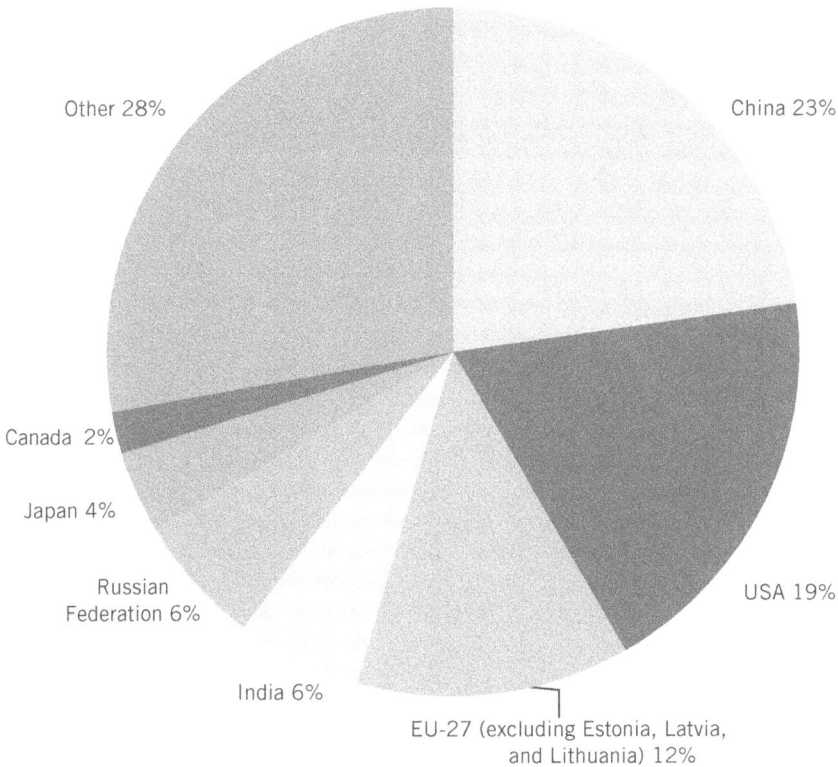

Other 28% China 23%

Canada 2%

Japan 4%

Russian
Federation 6%

USA 19%

India 6%

EU-27 (excluding Estonia, Latvia, and Lithuania) 12%

Figure 1.1 2008 global CO_2 emissions from fossil-fuel combustion, cement manufacturing, and gas flaring (million tonnes)
Source: EPA, 2013

in developing countries where they would represent the best solution for the country in question. In short, energy access for development should take priority over carbon-emission reduction (Nygaard, 2009).

The authors have adopted a similar 'development first' approach in this book, highlighting the benefits of low-carbon technologies where they represent the optimal energy technology for the countries or communities in question, from an overall economic, technical, socio-cultural, and environmental standpoint. It is the opinion of the authors that a 'low-carbon agenda', albeit essential from a global perspective, should not impede the drive to achieve universal access to modern energy services in developing countries as soon as possible. It should, however, be noted that a move towards an increased use of low-carbon energy sources often results in numerous additional benefits for a country and its population. These can include increased national energy security and balance of payments, and, in particular, local livelihood benefits, especially in rural areas where the local production of energy sources and their conversion to a range of energy services leads to significantly increased added value and income for the local communities (e.g. through fruit drying, jam making, and baking).

Challenges for delivering energy access

There are many reasons why energy access interventions have so often been unsuccessful; for example, the choice of technological equipment and its impact on the end users. The equipment may have been inappropriately designed; requiring spare parts that are hard to obtain in remote areas, or may simply have failed to meet the needs and cultural norms of its intended users. Its perceived value may not have justified the cost for the users; priced out of reach for the target market who also lack a supportive financial infrastructure.

Experience has shown that the main contributing factors that lead to the failure of energy delivery include capacity, finance, and governance. The task of increasing access to modern energy services is highly complex and fraught with a range of barriers, particularly if the intervention is trying to target remote, rural communities, and even more so if they are poor (Guerra-Garcia, 2004). Large distances, difficult terrain, limited purchasing power, and low projected levels of consumption by the end users generally make rural areas unattractive investment opportunities for both private-sector companies and government-funded utilities. These entities often prefer to concentrate their efforts on urban and peri-urban areas where there are greater numbers of consumers, and, from the perspective of politicians, potential voters (Haanyika, 2006). In poor and remote areas, which often have low levels of education and poorly developed infrastructure, new energy systems are often unable to access sufficiently high standards of local technical and managerial capacity, or set up supply chains for producing or sourcing replacement parts. In addition users often have relatively low capacity in effectively maintaining the technological system, and in setting up socially and culturally appropriate management systems to effectively oversee and finance the ongoing operations.

There is a lack of favourable institutional frameworks that encourage financing for potential energy business start-ups – commercial banks are often unwilling to lend to what is perceived to be a 'risky' sector – as well as a lack of support for the development of the sector through the establishment of complementary land rights, regulation, and subsidy policies (e.g. perverse incentives for fossil fuels). In addition, a poor or incomplete appreciation of the local socio-cultural norms (such as decision-making structures) and preferences (such as for one type of cooking practice or product over another) has often hindered energy practitioners from achieving the success they initially anticipated.These important lessons need to be captured and shared (Wilson et al., 2012).

Due to the complexity of effectively delivering energy services to the poor and underserved, private, public, and civil society organizations need to better coordinate and complement each other in trying to achieve the goal of universal access. While some profit-driven companies and social enterprises have started to make progress in the provision of a range of energy services and technologies to certain monetized sectors of the populace (e.g. the sale of solar lanterns or solar home systems through cash or credit to middle-income populations, highlighted in Chapter 5), both public and non-profit organizations also have important roles to play in bringing products and services to the poorest communities, households, and businesses. This role may include, for example, the dissemination of fairly large-scale, capital-intensive energy infrastructure systems in remote areas. Transportation access to these areas may be limited, and the intended beneficiaries unable to pay the full cost of the tariff repayments, including the sometimes high capital costs, such as for micro-hydro and biomass gasification systems.

The World Bank has described this as working beyond the affordability frontier, yet still operating within the boundaries of a market system: investing in projects that are not viable under normal market conditions unless the costs are, to some extent, subsidized. Often these systems require support to meet their up-front capital costs, but are then able to independently finance their ongoing maintenance and operating costs (Navas-Sabater et al., 2002). In doing so, interventions which have the potential to be highly equitable, by reaching a much wider sector of the population, and using technologies which have high initial capital costs but also high economic net benefits (not just financial benefits) can still be deployed. An example of this is renewable energy mini-grids which utilize a range of energy sources (micro-hydro, wind turbine, solar or biomass gasifier systems). Despite requiring high initial investment, these grids can provide energy for a wide range of productive uses, thereby generating significant livelihood and welfare improvements for the local community (see for example Yadoo, 2012; NRECA, 2010).

Aim of this book

The overall aim of this book is to define and analyse a range of energy delivery models that have been designed to reduce energy poverty in middle and low-income countries. It is hoped that by gaining a better understanding of the systems that affect these models, it will be possible to identify their successes and failures, and understand which are likely to be most effective at increasing energy access for poverty reduction. Drawing on a range of experiences from the field, and analysis of key aspects of energy delivery models, the book also aims to outline a framework that can guide energy practitioners in the design and implementation of their energy delivery models, hopefully leading to increased energy access and the reduction of energy poverty.

Chapter 2 ('Designing the delivery of energy') details what is meant by energy delivery models, including their historical development and categorization, and the factors that lead to their success or failure. It breaks down energy delivery models into their three main segments: the market chain, the supporting services, and the enabling environment; and describes the interdependent relationship between the actors in these three areas. It goes on to outline the categorization of energy delivery models into on-grid, mini-grid, and off-grid systems, as well as outlining the full range of stakeholders and actors who are involved in the delivery of each segment of an energy system.

Chapter 3 provides greater detail of on-grid energy delivery models, including relevant case studies, while Chapter 4 offers greater detail on mini-grid energy delivery models, and Chapter 5 provides details of off-grid energy delivery models. Finally Chapter 6 focuses on the main conclusions and recommendations that have been drawn from the analysis of a range of energy delivery models, the actors involved, and their interaction.

CHAPTER 2
Designing the delivery of energy

This chapter focuses on conceptualizing the delivery models that are used to provide the full range of energy services required by households, communities, and businesses. The emphasis is on energy for developmental purposes, particularly for the underserved urban, peri-urban, and rural poor in middle- and low-income countries.

Keywords: enabling environment, energy delivery model, energy distribution, participatory market mapping, supporting services

Defining energy delivery models

There is still no universally agreed definition of an 'energy delivery model,' despite its increasing use in literature relating to international development and public service delivery (Yadoo and Cruikshank, 2010; Renewable World, 2012; Wilson et al., 2012; Yadoo, 2012). However, the authors have created their own definition based on energy access research and their experience of practical energy delivery case studies. In particular, this definition has been informed by the use of participatory market mapping and business model design of a range of energy initiatives from developing countries, as well as analysis such as the development of the energy model tool detailed in Box 2.1 (Albu and Griffith, 2005; Sanchez, 2008; Wilson et al., 2008; Practical Action Consulting, 2011; Bloomfield, 2012).

Box 2.1 Energy delivery model tool

As part of the Department for International Development (DFID)-funded Policy Innovation Solutions for Clean Energy Research (PISCES) bioenergy research project (PISCES, n.d.), an analytical tool to assist planners and designers of energy access projects to better understand energy delivery models has been developed. This tool has, in turn, helped to develop the ideas used within this book and is freely available to practitioners online (Practical Action, n.d.). The tool highlights where particular combinations of energy sources, maintenance plans, management structures, and financing may either work together or be incompatible.

The user of the tool identifies key aspects of the delivery model, relating to the use of the energy, the energy supply, and whether the system is stand-alone, decentralized, centralized, etc. They then go through various aspects of ownership and management of the initiative as a whole, the conversion equipment, local resource rights, and the energy appliances used by the end users. On the basis of the information provided, the online programme then generates a customized market map, allowing project designers and practitioners to better understand the wider environment that might affect the success of their project.

---

http://dx.doi.org/10.3362/9781780447612.002

Energy delivery models are described as systems that include a set of processes and activities aimed at delivering energy services to end users through a market chain; a range of external services that support the good functioning of the market chain itself; and the overarching set of administrative and physical infrastructures that together with the socio-cultural context (Wilson et al., 2012) create the enabling or disabling environment for the market chain to function effectively. An energy delivery model often includes a number of stages from the initial locating and capturing of the energy sources, to their processing and supply, and distribution and marketing, including the technological infrastructure and the design and production of the appropriate technologies. It also includes the financing of the various segments and management and maintenance systems that ensure the energy is affordable and can be supplied over long time periods to meet a range of needs.

Each of *these processes and activities are carried out by* an individual, or a group of organizations, from local informal entrepreneurs to multinational companies (the market actors of the original concept), who work together to contribute isolated parts, or the full range, of activities to deliver energy services. These interrelationships are often highly complex, which is why it is so important to fully map out all the organizations involved and understand how they relate to, and influence, each other. This will be described in the next section.

Developing an energy market map

A market map is a diagrammatical method of representing a market system. The market mapping approach was first developed by Albu and Griffith (2005) in the context of Practical Action's work on agricultural market chains, and has now started to be used within the energy sector. A market system comprises three key segments: the market chain actors (those involved in implementation, including supply chain partners and contractors), the supporting services (including financial services, market data and capacity-building services), and the enabling environment (which includes regulatory and socio-cultural factors). Although each organization is important, equally important are the relationships between each and how they interact within the overall system represented by the map.

In an ideal situation the energy market maps are created through a process of participatory market mapping, including as complete a range of market chain actors and end users as possible. This helps to build market literacy and a greater understanding of the overall market system, including the relationships between each actor. The process of drawing up an energy market map helps to stimulate dialogue, reflection, awareness, and systemic thinking among the stakeholders, including the market chain actors, the supporting service organizations, and the policy makers. This can stimulate reflection on how to increase opportunities for poor producers and energy users. The participatory market mapping approach has proven effective in the analysis of bioenergy markets in the context of the PISCES energy access programme (e.g. Ndegwa, 2011; Bloomfield, 2012) and has been combined with the International Institute for Environment and Development's (IIED) delivery models framework in a practical guidance document developed with the official Catholic aid agency for England and Wales (CAFOD) (Bellanca and Garside, forthcoming).

An energy market map enables all the processes and organizations involved in delivery of energy services to be identified, including the *market chain*, from the initial collection of the energy source to its processing and distribution to the end user, together with the required energy conversion equipment and appliances. It also allows for the identification of the key factors of the *enabling environment* and the additional *support services* that need to be in place for the energy delivery model to function properly (see the energy market map in Figure 2.1).

The market mapping of energy systems can also contribute to national planning processes, particularly if policy makers are involved in the process, and can help energy project planners improve the delivery of existing energy systems or develop new, more comprehensive, energy delivery approaches. Detailed participatory market mapping has been carried out for a number of bioenergy systems as part of the PISCES research programme, with workshops conducted in Kenya and Sri Lanka in 2009. The workshops sought to identify key actors in each bioenergy market system, and to understand how they are linked to each other, as well as the factors affecting them and blockages that needed to be overcome (Ndegwa, 2011). Similarities were noted in the bioenergy market systems of both countries and an analysis of the maps identified how specific actors from each system could tackle particular barriers and issues in order to improve the efficiency of the overall energy systems.

The energy market map enables practitioners to gain a greater understanding of the functioning of often complex energy delivery systems, by breaking them down into their various components. For the purposes of this book, the authors have adapted the original market map in order to better assist in identifying ways to support the delivery of energy to the poor.

The systematic process of analysing delivery models is important for energy practitioners as they continue to try to design effective approaches to delivering a range of energy services either by improving existing systems, or through establishing new energy businesses. By identifying the key segments of an energy market map, including the organizations and actors that are responsible for delivering each part of an energy system, it is possible to identify which aspects are particularly important for ensuring increased access to energy services for the poor. Overlooking a single element of the market map, might lead to the failure of the delivery model; conversely, being aware of the role of a particular element might enable a successful outcome.

Enabling environment

- Skills and capacities to manage and maintain systems
- Local acceptance of technologies, approaches
- Local norms and behaviours
- Willingness and ability to pay
- Local civil society organizations and structures
- Level(s) of conflict and criminality
- Access to markets for the poor
- Local awareness and energy-use practices
- Institutional strength
- Global/regional geopolitical trends (resource availability, prices, regulation requirements)
- State of country's infrastructure (roads, grid coverage)
- Environmental conditions
- Tax and tariff regimes
- Government policy, law, strategies
- Funding for R&D programmes
- Resource rights and allocations

Energy market chain

Energy resource → Production / processing of resource → Production/distribution of appliances

Ownership, management, M&E — Technology installation, maintenance → Distribution of energy → End use (usage, payment systems)

Supporting services

- Capacity building
- Technical support and services
- Energy technology and appliance R&D
- Financing opportunities
- Knowledge management – database, networks

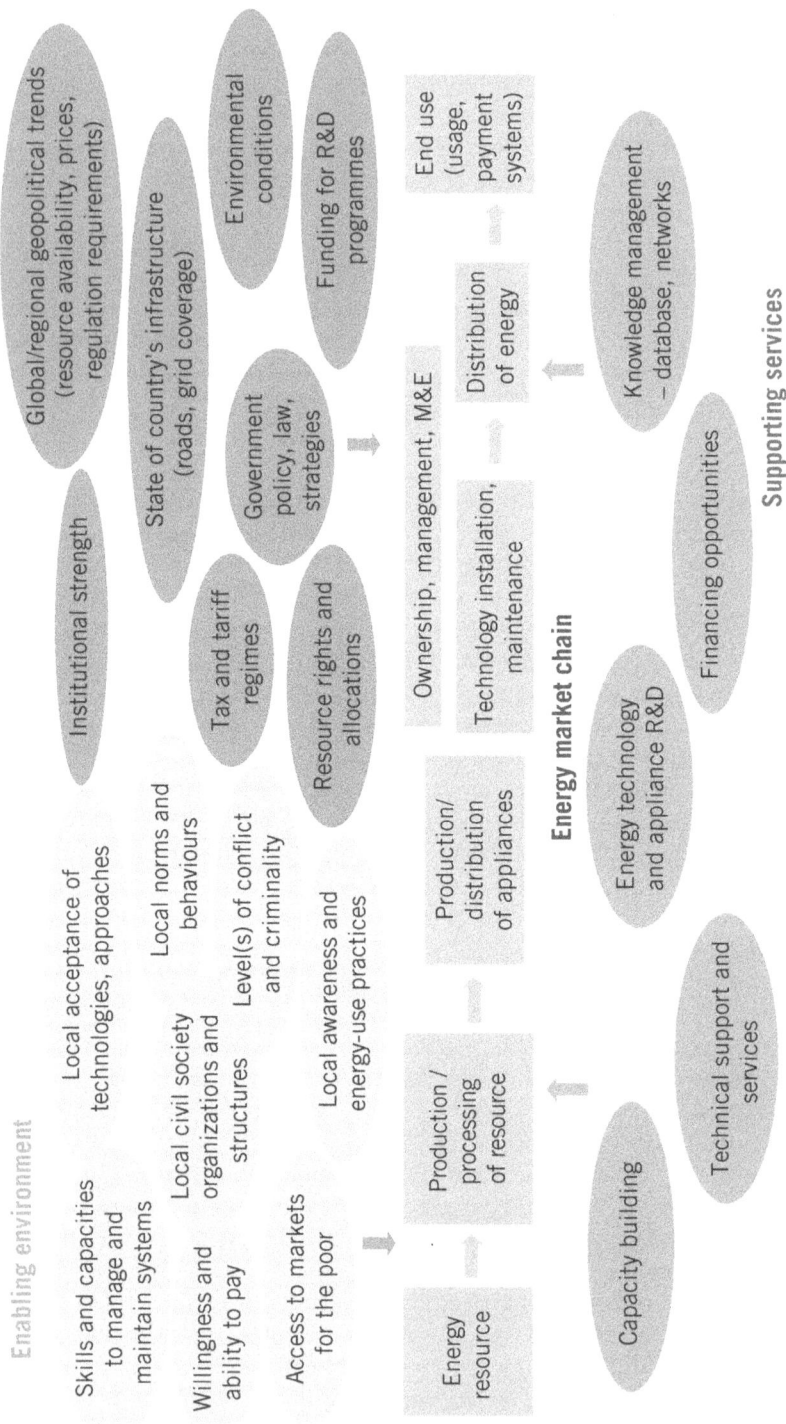

Figure 2.1 A generalized energy market map highlighting the three key segments

The authors have further categorized energy delivery models according to their methods of distribution into on-grid, mini-grid, and off-grid, and developed energy market maps for each of these. It is important to state that this categorization does not only refer to the distribution of electricity, but also to all types of energy services or products, which can be distributed through a centralized grid, a mini-grid, or packaged separately from a grid.

The success of energy delivery models

Before going into the details of specific energy delivery models, it is important to understand that increasing access to energy is not only about transferring a range of appropriate technologies and services, but also about understanding the most appropriate segments within each delivery model including their ongoing management and maintenance, as well as ensuring the needs of the end consumers are met sustainably. Energy delivery models need to be replicable and scalable if they are to result in a broader development benefit beyond the pilot or experimental stage. An analysis of appraisals from a range of energy interventions over the years has identified different, and sometimes divergent, definitions of success. Box 2.2 outlines a set of key factors for success in an ideal energy delivery model, compiled by the authors.

Box 2.2 What is success?

1. The delivery model is designed to be financially sustainable in the long term – i.e. once implemented, the energy solution can continue to pay for itself throughout its lifetime.
2. The delivery model reaches an appropriate *technical standard* – it fulfils its technical potential to produce appropriate energy services that meet the needs of the end users, and is well designed and maintained.
3. The delivery model has *a positive social impact* – it creates sustainable welfare and livelihood benefits at the local level, as well as the provision of energy.
4. The delivery model has a *positive*, or at least neutral, *environmental impact*, at the regional and global level.
5. The delivery model is *replicable* – it can be adapted to different contexts or scaled up to reach more beneficiaries, in other areas, and ideally other countries and regions.

The principal and fundamental aspects that determine the success of any energy delivery model is how suitable and compatible the energy solution is to the community as a whole or to individual end users. Analysis of a wide range of energy delivery models from around the world has led to the identification of five key factors which are believed to be pivotal in shaping the successful delivery of energy products and services: social, environmental, technological, financial, economic, and environmental. The way that these factors are considered in the initial design, combined with the ongoing management and maintenance of the systems, and the interactions between all the involved actors, will ultimately determine the level of success of an intervention in reaching the poor and underserved.

The ways that each of these five factors helps to ensure the success of an energy delivery model is outlined here:

- *Social factors* comprise the social and cultural characteristic norms of the specific delivery model's target end users as well as the market actors themselves. These include the traditional decision-making structures, particularly in remote communities, which are often not understood or fully appreciated by people from other countries or even by country-level policy makers. These factors also include the level of trust between the implementing entity and local communities and decision makers; local perceptions of a range of technology options, based on established energy-related practices and preferences; health issues and local perceptions of these and how they are prioritized; and local capacities for management and maintenance of an energy programme or initiative, particularly concerning locally shared resources (such as rivers or forests for hydro and bioenergy resources). They determine the long-term impacts of energy programmes such as the improved health, education, and livelihoods of users.
- *Environmental factors* include the negative impacts on natural resources or their reduction over time, such as decreasing output from hydro facilities due to changing rain patterns, wind power, and solar radiation levels. Cases of negative impact include: issues related to waste management of new technology components, such as batteries for solar systems; hydro schemes which flood large areas of environmentally sensitive land or have a negative impact on local flora and fauna; and biofuel projects which replace prime forests with crops that destroy biodiversity.
- *Technological factors* refer to the availability and suitability of appropriate technologies, such as well-designed improved cook stoves which are efficient, suited to household cooking techniques and pot sizes, as well as being well-liked, as opposed to poorly designed stoves that do not meet the end users' needs and are therefore not accepted; solar home systems which can be sourced and maintained locally and are suited to the particular climatic conditions; and wind turbines which function efficiently in remote and extreme situations such as mountain regions. Similarly important are the technical expertise requirements for maintenance and/or the availability of spare parts.
- *Financial factors* relate to the impact of economic processes, such as subsidies, and their potential to distort the market; and the cost of producing the energy and its affordability to the end users, particularly relevant for renewables where investments can be high. For example,

a solar home system might only have a very small range of financial options that will be acceptable to the users (which will result in only a very small part of the financial circle, in Figure 2.2, being shaded).

- *Economic factors* relate not to direct financial benefits, but more indirect impacts of the energy system on individual households or society as a whole. This includes the potential economic benefit of improving the health of households through reducing the emissions of cook stoves; of saving forests that would otherwise be destroyed by charcoal production; and of providing households with extended, potentially productive, evening hours through supplying a good quality light source.

Once the conditions imposed by the social, environmental, technological, financial, and economic, factors have been considered, only a much narrower range of options will be fully acceptable by all target users, as highlighted in Figure 2.2. This figure represents a simple schematic of each success factor. The overall success of an intervention only occurs within the centrally shaded area, where all the success factors overlap. It is important to note that while compatibility issues can be addressed through a process of adaptation or adjustment following an intervention, the long-term success is likely to be considerably higher if all potential success factors and barriers are suitably considered and addressed during the initial design and planning of a new energy intervention.

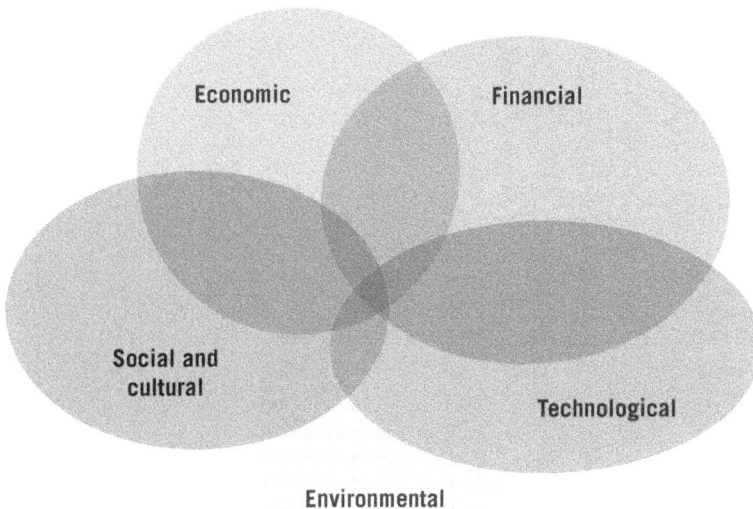

Figure 2.2 Factors that determine the success of energy delivery models

Often some of the potential success factors, or barriers to success, may not be understood, or even be of concern to the energy users themselves, who often focus only on the short-term outputs and issues that directly affect them. This is generally the case for public health and environmental issues, such as the overall health benefits of a particular stove programme, or the resulting problems of deforestation and soil degradation, as these are generally expected to be the concern of policy and decision makers rather than individual households. These environmental and economic factors are very important in determining the overall success of an intervention, particularly if financed by development funds, where programme designers expect to achieve specific impacts which may not necessarily be aligned to the priorities of the energy users. On the other hand, social and financial issues are more likely to be primarily the concern of the users, for example in the case of traditions that might impede the adoption of a new technology, such as using ethanol as a household cooking fuel. This particular issue has in fact already started causing problems relating to alcoholism in Kenya and Ethiopia, and so is being tightly controlled by their governments (direct experience from Practical Action's work in western Kenya). Increased and more widespread production of ethanol is resulting in increased drinking of ethanol. However this can be overcome through tight controls to ensure the ethanol is denatured at source and therefore undrinkable.

It is important that the funders and related policy makers for energy interventions are fully aware of the social and cultural impacts of the energy delivery projects they are supporting. It is key that they do not only focus on pushing ahead their own agendas, as this will significantly impact on long-term success of projects. In other words, it is essential for energy practitioners to understand the interaction and interdependence of all the relevant factors in order to implement a successful energy delivery model.

Previous categorization of energy delivery systems

An in-depth understanding of energy delivery interventions and processes through their conceptualization, classification, and rating gives energy practitioners a clearer idea of the failures and successes of such interventions. This understanding can then demonstrate how workable models could be replicated to reach higher levels of scale within a local or global context to significantly increase energy access. Over time, there have been successive changes related to the way energy delivery models have been defined and designed, with a general evolution from a predominantly technology-focused model to more holistic approaches. This gradual change has included the increase in recognition of the importance of the end-user perspective at the household and community level, with a focus on the long-term viability, sustainability, and efficiency of energy delivery (Best, 2011; Wilson et al., 2012; Yadoo, 2012). Various considerations include participation, ownership, inclusion of different groups (particularly women), efficient and productive uses, financing, and other factors (Sanchez, 2010). However, despite this evolution involving a wide variety of actors and approaches, energy access still remains a significant issue in many parts of the world as highlighted in Chapter 1.

In the past, energy delivery models have been analysed and categorized in different ways according to a range of criteria, with their categorization often being based on their scale, impact goals, and sources of funding, rather than the set of measurable and comparable functions making up their internal structure and organizational relationships. Box 2.3 highlights the classification of projects by implementation scale: distinguishing between international, national or local level. The classification does not differentiate between the types of technologies involved, the financing arrangements or the energy resources.

Box 2.3 Classification by implementation scale

International-level energy delivery

International models of energy delivery have been mostly linked to aid and international policies in order to achieve significant and long-term goals on energy access and development. For example, the International Monetary Fund (IMF) and the World Bank developed the *Poverty Reduction Strategy Paper* (PRSP) in 1999. This paper included plans for the development of global energy delivery models in low-income countries, with bilateral and multilateral aid focused on participatory planning and implementation. It must be noted that there were varying levels of energy delivery planning within the PRSPs in different countries, and while this provided a good opportunity for energy practitioners to try to follow a development-first approach, the energy sector was still dominated by a 'grid connect' approach, often favouring the urban or well-connected geographical regions.

National-level energy delivery

National-level programmes have traditionally been designed by governments, but often either partially or wholly implemented by third-party organizations, often through concessions and agreements with utilities. These programmes are generally tied to government plans and programmes, with their implementation often being centrally planned and executed, and through top-down approaches. An example of a national-level energy programme is the rural electrification programme in China, which aims to achieve full electrification for rural people by 2015 (see example in Chapter 3).

Local-level energy delivery

Local-level energy delivery focuses on providing energy services to specific groups of people within geographic boundaries, often to specific numbers of users (an individual, community or village). These energy services are also referred to as 'small-scale' or 'community' projects and programmes, and have often been tied to international and philanthropic aid. The entry point has generally been to attend to certain specific energy needs of the target group, such as electricity, cooking, lighting, healthcare facilities, lighting for schools, etc. Among the most known models developed and tested are the energy service company (ESCO), energy contract company (ECC), small enterprise management (SEM), and renewable energy service company (RESCO) models.

Another approach in categorizing energy interventions has been that of grouping delivery models together depending on the type of organization responsible for their implementation. This has generally included national or local government, private sector, international donors or development banks, or a combination of a number of these. The *Pró-Álcool* Ethanol Programme in Brazil represents an example of an energy intervention predominantly implemented by a national government; the Solar Electric Light Company Programme in Sri Lanka an example of a private-sector led energy intervention; while the Rural Energy Enterprise Development Programme in Bangladesh represented an intervention led by an international aid organization. Trying to categorize energy interventions using this approach, is clearly too simplistic, and does not allow the reasons behind the weaknesses in the delivery of energy services to be uncovered.

In addition, the identification of the main actors involved in the delivery of energy systems, and the various roles that they are able to play and interact, is of great importance. A list of the main actors that are involved in the delivery of energy services is provided in the 'Actors delivering energy' section of this chapter.

Energy delivery models have also been classified with a focus on their type of management, both public and private, as described in Box 2.4. This classification does not differentiate between the different energy sources, and their processing and packaging, or the role of the actors that might influence each system.

Box 2.4 Management-based classification

Decentralized virtual utilities: Operating on the principle of 'fee for service', this model consists of the implementation of small-scale systems in the premises of the users and charges monthly fixed payments. The ownership and responsibility for running the systems is given to energy service companies (ESCOs), entities created ad hoc for this purpose.

Micro, small, and medium-scale enterprises (MSMEs): Many of these are based on private ownership and management systems, including:

Local electricity retailers. Small business or cooperatives establish an electricity retail business and the community makes its own arrangements among its members to run the energy services. The common feature is the ability of retailers to prepare sound business plans to obtain credit financing or backing from stronger partners.

Energy equipment producers and retailers. Small-scale technologies are produced and/ or distributed through small local dealer networks to penetrate isolated rural areas. This model makes use of the low cost of local labour and local knowledge of the different social, economic, and cultural issues, in order to sell energy equipment. One of the delivery mechanisms within this model is the creation of the financing structure that enables the dealers to extend credit to low-income families. However there are high risk factors involved, primarily due to potential delays in payment by the customers.

Energy concessions: The government gives an area or region as a concession through a competitive bid to provide electricity to rural families with the winner signing a concession contract of electricity services through the lowest cost option. The energy provision is supervised by the regulator and subject to quality and quantity requirements. This model has shown potential for effectiveness in service delivery, but could be quite expensive for the state as it generally requires high subsidies in order to stretch the services to poor markets such as isolated rural areas (this is explored in more detail in Chapter 3).

While each of these methods of categorizing energy delivery models has its merits, they are all too simplistic, and are not able to capture the full complexity of delivering a range of energy services, particularly to the poor and underserved. The authors have integrated many of the categorization methods mentioned in this section to develop a more comprehensive framework for outlining and defining all aspects of energy delivery models.

Actors delivering energy

The process of delivering a range of energy services and products to end users is dependent on a range of actors who play direct or indirect roles in one or more segments of an energy delivery model. Sometimes an organization can play more than one role and act as both a market chain actor as well as a supporting service entity depending on the circumstances. These actors range from non-governmental organizations (NGOs), to companies and individuals, cooperatives, communities and governments, each with unique roles depending on the type of energy and the type of delivery.

There are no strict boundaries on what the involvement of different actors can be, but Figure 2.3 highlights some of the broad trends.

The following section outlines some of the individual roles of the main actors involved in delivering energy services, emphasising the most important roles they each play, and the types of interventions that are best formulated and implemented by partnerships between them.

Governments. Governments play the most important role in creating the right enabling environment and providing supporting services for energy delivery models. Central governments can also choose to play a more major role in the market chain process by actually implementing projects and programmes, either at the regional or national scale. At the local level, such programmes can be implemented by local government institutions who lead on their design, fund management, and/or implementation. For rural electrification, governments in both developing and emerging economies continue to play a central role. A common characteristic of all government energy access programmes is that they are often subsidized, and this funding capacity is frequently identified as one of the main reasons for their success. However it can also be detrimental if it leads to market distortions that are later difficult to remove (more nuanced discussions on the role of subsidies for each scale of energy delivery are given in Chapters 3, 4, and 5).

Government-led programmes may be 'direct' implementation top-down delivery systems, usually centrally designed, implemented, and managed by government (often without the participation of the end users and with limited consultation). Alternatively, they may be 'indirect' implementation, where the central government delegates regional, provincial, and local bodies to coordinate, plan, and deliver the services while it promotes the services, establishes the rules and regulations, and provides funding.

ACTORS / ROLES	International agreement	National policy formulation	Regulation, tax and incentives	Resource assessments	Project/initiative design	Grant funding	Commercial financing	R&D/technology development	Technical assistance	Loan guarantees	Construction	Product/service distribution	Microfinance provision	Provision of feed stocks/fuel	Marketing	Operation	Service purchase/lease	Maintenance
International bodies	O					P			P									
National government	O	O	O	P	P		P		P									
Local government					P	P												
National utilities				P	P			P				O			P	O		
Banks/financial institutions							O			P								
International donors					P	O			P	P								
Technical experts				O	P			P	O									
Large private sector					P		P				P	O		O	O	P		
Small-scale entrepreneurs					O						P	P			P	P		O
Agriculture and forest sector														O				
Microfinance institutions					P								O					
Universities/R&D								O										
NGOs					O	P			P		P				O	P		P
Cooperatives					P											P	P	P
CBOs					P											P	P	P
Consumers/households																	O	P

Legend: ■ Often undertake   ▨ Potentially undertake

Figure 2.3 Roles and responsibilities of different actors for energy delivery
Source: Practical Action, 2010

The private sector. These actors may be companies or individual entrepreneurs managing energy utility companies (large or small), or businesses that directly or indirectly sell products or services to customers (both formal and informal). Their operations occur at a range of scales and locations, and for different population groups. An increasing number of companies involved in delivering energy services to the poorest, particularly mini-grid and off-grid, function as social enterprises, led by 'for low profit' social entrepreneurs, who recognize the strong social benefits associated with energy delivery and are ready to compromise profit opportunities to take advantage of the social impact of their activities. Social entrepreneurship has developed as a way of trying to overcome the barriers to delivering energy to the very poor and isolated, who are often outside traditional markets and the geographic operations of most formal private-sector companies. It must be noted that donor, and/or government support is often also provided to private-sector organizations to allow them to deliver energy products or services to their respective markets.

NGOs and community-based organizations (CBOs). Aid agencies have been engaging directly and/or indirectly in the provision of energy access for the poor at a significant scale over the last few decades. Projects have often been funded by international aid agencies and implemented by independent organizations such as NGOs, church groups, local cooperatives, local government, and consumer associations (e.g. Practical Action, SNV, Christian Aid, and the German Agency for International Cooperation's Energising Development (EnDev) programme). Direct activities are often carried out in agreement with governments and vary in scope from national down to individual village level. An example of what has been largely reported as a successful dissemination project implemented with a mixture of private intervention and subsidies is the biogas initiative in Nepal (see Box 5.9 in Chapter 5).

Communities have also been engaged in project implementation in different ways, sometimes with the support of organizations that act as facilitators and donors. Community implementation is frequently focused on mini-grid and off-grid systems related to small-scale technology to supply local needs. As an example, community organizations have been involved in supplying agricultural inputs for energy production (e.g. farmer cooperatives producing *jatropha*, or producing biomass from agricultural residues such as rice husks or corn cobs). In some cases, local groups have engaged in the production of energy appliances (potters' cooperatives making improved clay cook stoves, such as the *Upesi* stove in Kenya, and the *Anagi* household stove in Sri Lanka – see Box 5.8 in Chapter 5); their marketing, promotion, and distribution; or have taken responsibility for managing their supply.

Development banks. These banks are often commissioned by governments to support energy access in the less-developed nations. This includes the World Bank, as well as the African Development Bank, Asian Development Bank, the Inter-American Development Bank, and the East African Development Bank.

For instance, the World Bank has been one of the most important promoters of small and medium-sized enterprises based on private entrepreneurship. Well-known examples can be found in off-grid rural electrification programmes in Nicaragua, Peru, and India. Development banks play a stronger role in the support of an enabling environment and provide services such as financing, technical advice, and capacity building.

Donors. National aid agencies (e.g. the German Federal Ministry for Economic Cooperation and Development – BMZ, the UK Department for International Development – DFID, the US Aid Agency – USAID, the Swedish Institute for Development Cooperation – SIDA, and the Canadian International Development Agency – CIDA), and other international agencies (e.g. the EU, the UN Development Programme, and the UN Environment Programme) play a major role in the delivery of energy services in developing countries. Donors are crucial in strengthening the enabling environment and providing supporting services rather than playing a major role in the actual market chain. Without donor support, many initiatives, especially those which focus on the poor or low-carbon and clean energy resources, may be struggling to effectively reach their target market. Donors may be in a better position, compared with governments or the private sector, to fund risky, pioneering technologies and approaches which may not become self-sustaining for many years. They often play significant roles in technical assistance and training, investing in research and system design, and often the capital costs of installations.

One of the biggest roles for donors, apart from funding, is to facilitate the collaboration of a wide range of partnerships, including those between different stakeholders, both public and private. These partnerships can then support innovative approaches, mechanisms, and collaborations providing grants to cover their initial costs, training, or take-off stage concepts. Donors can also play a significant role in research and development (R&D), particularly in innovative mechanisms and delivery services, thereby reducing the risk to the private sector in trying to develop these themselves. In Sri Lanka, it took two decades of product development and testing of improved cook stoves, supported largely by donor and government funds, before the final commercialization of the *Anagi* household stove which has now reached 3 million households. The details of this stove are highlighted in Box 5.8, Chapter 5 (Rai and McDonald, 2009; Stevens, 2010).

Financing institutions. These can be both formal and informal organizations. Rotating savings and credit associations are informal and provide credit in addition to savings. Many energy enterprises, especially in rural areas, are based around informal financing arrangements, and therefore access is limited or localized. Borrowing from relatives and friends is common, with no proper accounting required, and nominal interest or fees. Because of the nature of informality, lenders can end up losing money if receivers fail to make their contributions, and the available capital is generally limited, which makes the scaling up of efforts practically unachievable in many informal economies.

Formal institutions such as banks or microfinance institutions (MFIs), for example FINCA and Opportunity Bank in Uganda, are the common sources of financing but have stringent requirements that many energy project developers or suppliers may not be able to meet. Savings and credit cooperatives, or SACCOs (financial institutions formed between groups of individuals within a community who are able to lend to each other through pooling resources, such as *Stima* SACCO in Kenya), generally have quite informal structures. Many rural households or small businesses are served by MFIs or SACCOs which have fewer joining requirements. By using unconventional forms of collateral, such as group guarantee, they can provide loans that are unsecured or not backed by physical collateral. However many MFIs do not have specific energy portfolios and there is a need to build capacity for these institutions to provide loans to energy entrepreneurs or for purchasing energy products by consumers or businesses (although this is starting to change, particularly in East Africa through new programmes such as CleanStart in Uganda, and the EU-financed Sustainable Energy Markets Acceleration – SEMA project in Kenya, Uganda, and Tanzania).

Key segments of the energy delivery model

This section will provide greater detail on the three key segments that constitute the analytical framework of energy delivery models which can be analysed through the market mapping approach introduced in the Chapter 2 section, 'The success of energy delivery models'. These three key complementary segments are:

1. Energy market chain, comprising the actors who deliver the resources, technologies, processes of production and distribution to the end users.
2. Enabling environment, which constitutes the structures, regulations and incentives that support energy delivery at the government and policy level. In addition, it also includes global trends, the country's infrastructure, and aspects of the socio-cultural context of norms and practices.
3. Supporting services, which help catalyse the actors to act upon and provide better services and products overcoming barriers in the market chain, as well as creating a favourable enabling environment.

Energy market chain

The energy market chain refers to all the technologies, systems, and activities that are carried out by a series of market actors who, depending on circumstances, might cover a variety of roles from primary producers to distributors who reach the customers. The market chain lies at the heart of a delivery model and is the area over which practitioners potentially have the greatest influence due to their direct involvement. This chain covers access to resources, technology design and installations, conversion and processing, distribution, marketing, and final end-use energy appliances.

Resources, both renewable (biomass, wind, sun, water, geothermal, etc.) or non-renewable (coal, oil, natural gas, etc.) determine the technologies employed and the processes within the market chain. The power produced can be used directly but is more often processed and packaged, and used through a range of conversion equipment (as outlined in Table 1.3). In recent years, more innovation within the market chain has been initiated through the development of new technologies and processes to meet a range of energy needs, from cooking and lighting to processing of agricultural products (especially for rural consumers). It has been estimated that the overall proportion of renewable capacity in the world now comprises about a quarter of total global power-generating capacity, and supplies close to 20 per cent of global electricity, with most of this provided by hydropower (REN21, 2011).

Including the governance, management, and ownership structure of energy delivery, the market chain involves a range of market actors who belong to the private sector, government, NGOs, community organizations, and development banks. Particularly important for pro-poor energy delivery models in fairly remote areas are the distribution channels and payment structures in order to overcome the long distances, poor infrastructure, and low purchasing power of the end users. The final customers are also an essential part of the model as they ultimately determine whether the model design is appropriate and therefore its success. The analyses of these segments in the market chain will allow a clearer understanding of the various functions and processes needed to create reliable and efficient markets. Additional efforts are needed to provide services to those that are underserved, and product development may require more in-depth field research and participatory approaches (Wilson et al., 2012).

Enabling environment

The enabling environment includes the nature of governance within a country (national leadership, levels of civil society freedom, and ease of doing business) along with relevant institutional structures, government policies and regulations, resource rights and land tenure, and the state of infrastructure development. It also considers wider global trends, interests, and political configurations, where these are relevant to the delivery model. This might include a country's participation in climate change negotiations and how it develops its own climate policy, as well as the nature of its energy trade relations with other countries.

In the context of this book the enabling environment also includes all aspects of the socio-cultural context that are critical to effectively delivering energy to the end users, particularly the very poor and marginalized. Relevant aspects might include the nature of local leadership and social organization, local cultural preferences for particular products and services, and local willingness to pay for goods and services, based on perceived value.

The purpose of charting the enabling environment is to understand the dynamics that are affecting the entire market chain process, and to examine the powers and interests that are driving change. The enabling environment is almost always beyond the immediate direct control of the market chain actors, but may be amenable to change through a range of interventions, including awareness-raising, lobbying, or via development interventions at a national, regional, or global level.

Government policies

At a basic level, the political economy of the country, where an energy delivery model is being implemented, may be critical to its success. Models that can be implemented in a certain political context, with particular levels of support for a market economy or where governments have a strong control over rural development, may be impossible to implement in another country context. In China, for example, the government's top-down approach to development allows for the mass roll-out of rural development programmes and has had a positive effect on the uptake of decentralized energy programmes, such as clean biomass gasifier cook stoves and biogas programmes. In other countries, such as South Africa, where there is a strong government focus on on-grid expansion and very little support for decentralized initiatives, this hampers the ability of rural areas to implement decentralized electrification programmes. Where the government supports strong, independent markets, there may be a vibrant entrepreneurial class that is able to innovate and develop new approaches to energy access.

The policy and regulatory side of the enabling environment includes: economic and energy policies and laws; trading and quality standards and rights of access to natural resources (including land); consumer trends; global and national energy prices and tendencies; business regulation; tax and tariff regimes; tax exemptions; and government incentives. The level of transparency in the administration of public and private affairs, as well as the official institutions tasked with tackling corruption, plays a big role in creating a positive enabling environment. Government policies have a major influence on the levels of development of the country's infrastructure system (transport, communications, electric grid coverage) as well as levels of support for decentralized energy access, rural electrification, and so on.

In reaching the poor and underserved, various partnerships or interactions in the market chain are forged such as public–private and private-civil society, entrepreneurs, and financing institutions. For such partnerships to thrive, appropriate institutional and legal support is important; for example the exi-stence of effective laws governing contracts influencing public–private sector initiatives or those between communities, government, and the private sector. It is quite common for opportunities to be missed to realize mini-grid systems.

This is due to the lack of suitable frameworks for obtaining licensing to supply communities directly, rather than having to go through the national energy regulator to receive energy supply (Modi et al., 2006). Relevant government policies are also important for other off-grid energy services, including energy for cooking, for electrification and ICTs, and for mechanical power, which are important for increasing household energy access and for improving lives and productive uses of energy. This is highlighted further in Box 2.5.

Box 2.5 Policy support for different types of energy services

Energy access is considered a fundamental enabler for achieving the Millennium Development Goals (MDGs). Modi et al. (2006) describe at least three types of energy services required to fulfil the MDGs and the types of policy support they need:

- **Energy for cooking:** Policies are required to enable the use of modern fuels rather than biomass and/or adoption of improved biomass cooking stoves. Fuel switching policies, such as direct subsidies or lease/finance mechanisms and bank loans for lowering up-front costs, need to be designed. In the biomass sector there is the need to design targets, especially for the scaling of improved cook stove programmes and sustainable biomass production.
- **Electricity for illumination, ICT, and appliances:** Policies are required to support household and commercial activities and the provision of social services, to ensure reliable access to electricity in all urban and peri-urban areas. Successful policy and regulatory support can help lower the cost of initial connections or spread its cost over time, especially for poor households. Such support can also assist small independent power producers to operate under a regulated environment; can help governments to ensure that utilities are able to recover recurring costs of generating electricity to protect financial viability of investment; and can assist with the regularization of tenure for slum dwellers to increase market size and formalize electricity demand. Tunisia, for example, developed a national policy objective of rural electrification for all, as a minimum standard for public service with a goal of providing 100% electrification by 2010.
- **Provision of mechanical power:** Policies are required to support the deployment of me-chanical power systems for operating agricultural and food processing equipment, to carry out supplementary irrigation, to support enterprises and other productive uses, and to transport goods and people.

Similarly, the requirements for generating private-sector involvement may include: measures for full disclosure, particularly of liabilities; access to credit information to allow investors to assess risks; a requirement for clear banking procedures for financing enterprises; and the provision of underlying guarantees for renewable energy developers to attract investment from banks and other financial institutions.

In addition, the importance of clarity on property and land tenure rights is of critical importance, with compensatory provision being required if land use is involved. Clearly defined standards for energy products are important for ensuring quality control and that users have access to information about the products they are purchasing. Energy, land and resource rights, and ownership (particularly for bioenergy), as well as business registration, are very important for the installation and management of certain energy facilities and systems. Regulations are also key to protecting consumers, as the installation of energy infrastructure can lead to the formation of monopolies, whether in public or private hands, and the temptation always exists to abuse this position.

Figure 2.4 shows the enabling environment factors that affected 15 small-scale bioenergy case studies, from a 2009 joint Food and Agriculture Organization (FAO) and Practical Action study, which includes a summary of consumer trends, global business regulations, tax and tariff regimes, tax trade standards, and official corruption quality assurance institutions. This figure highlights the relative importance of particular enabling environment factors, with national government policies, regulations and standards, and financing being the most important.

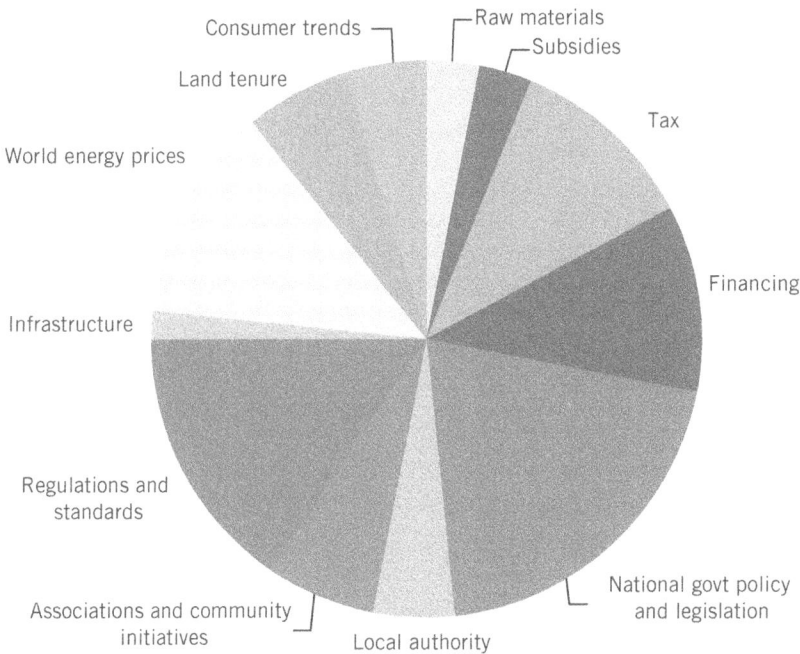

Figure 2.4 Factors affecting enabling environments in 15 small-scale bioenergy case studies
Source: Practical Action Consulting, 2009b

Socio-cultural context

The socio-cultural context is critical in ensuring that energy delivery models are designed to meet the needs of the poorest end users, as well as for assessing their success and potential for replication and scaling up in other areas and contexts. Given its key role, aspects of the socio-cultural context have been emphasized within the enabling environment discussions, as well as in the energy delivery market maps (Wilson et al., 2012). The socio-cultural context takes into account social and cultural norms, such as traditional decision-making structures, established energy-related practices and preferences (e.g. for wood-burning stoves rather than solar cookers) and gender-related preferences and limitations as well as local capacities for management and maintenance of an energy programme or initiative. Broader societal influences include consumer trends, political pressures, and governance practices (Wilson et al., 2012).

However, it is important to note that there is no blueprint for developing inclusive energy delivery model design as the socio-cultural context varies widely from country to country and within countries. A great degree of community ownership and participation has been key to the sustainability of micro-hydro projects in Nepal (Yadoo, 2012), but has not been the case in similar projects in China, which are more centrally controlled (Clancy et al., 2004). The uptake of cook stoves depends on cultural acceptability and taste preferences, and therefore benefits from the high levels of local participation in design, production, and distribution (Wilson et al., 2012). The attention to such critical aspects in any delivery model will ensure greater responsiveness to local needs and preferences.

As the goal of an energy delivery model is to provide a range of energy services to end users, it is logical that market research should include local consultation with end users and market chain actors, to ensure the socio-cultural context is understood and taken into consideration in model design. Business-orientated development initiatives have demonstrated that participatory market research greatly enhances the quality of information that is gathered and can help in the design of locally appropriate products and services. The literature also suggests that stoves designed in laboratories without fieldwork experimentation are less readily adopted than those designed with the local users to specifically meet their preferences (Barnes et al., 1994; Cecelski, 2004). In some cases, cook stoves have been rejected on cultural grounds as their designs have not taken into account the cooking styles and cultural preferences of the users (Vermeulen, 2001). Clancy et al. (2004) suggest that improved cook stoves may have been taken up by a higher proportion of households in China because women are directly engaged in the cash economy, and therefore have a clear interest in time saving. Such stoves can save time, even if they require an adjustment in some aspects of cooking practices. It is important that energy delivery models understand the socio-cultural context and the importance of a range of factors that users respond to when making decisions about using a new technology, including time saving, stove cleanliness and emissions, and price.

It is also important to note that although an understanding of the local context is important, this is different from adopting a community-led approach. In community-led participatory models, the increased awareness of community needs and perspectives does not necessarily equate to increased participation. One of the factors that can lead to poor participation is that communities can be fractious and inequitable, with the natural leaders often taking control. In many developing countries, men are used to taking the lead in official discussions with women taking a more subservient role, and so often the information received by energy practitioners, particularly concerning household cooking, is not always gender representative. Community cohesiveness and organization, particularly concerning household cooking, is often varied, and therefore participatory approaches need to be specially designed to ensure that the voices of all end users are heard. Factors such as the local level of community organization, cohesion, culture, education, average income, and income variation, as well as the presence or absence of key individuals, all have an impact on whether a participatory model will work as well as anticipated. Most communities have unique power relations, significantly when it comes to gender and vulnerable groups, and therefore there is no guarantee that a community-led approach will bring more equitable outcomes. It is often hardest for the poorest people to participate, due to a need to focus on basic needs and shorter-term provisions. However, the potential for the improvement in livelihoods, through increased access to energy services, is such that significant efforts to include marginalized groups will almost certainly represent a good investment of development funds.

Another important aspect of the socio-cultural context concerns user expectations of modern energy services. While many end users would like access to affordable grid electricity, they are not necessarily willing to take on the increased burdens and responsibilities that this implies, particularly meeting the initial connection fee and the ongoing regular payments. Households that have purchased small solar home systems are often disappointed to find out that the level of electricity provided is not able to fully meet their expectations, such as for powering televisions and refrigerators (Best, 2011). It is therefore important that the actors delivering energy are clear to their consumers concerning the quality and quantity of the energy services they are selling.

Above all, there is no blueprint for successful interventions, because success factors are context specific. Thus, it is important to find out what success looks like from the perspective of the target users, or communities. There is a general consensus that successful projects are the ones that benefit local people and which endure. The prioritization of global issues such as indoor air pollution, deforestation, and carbon emissions may not necessarily align with the concerns of the poor and can detract from the immediate issue of energy access (Zereffi and Wilson, 2010). A good understanding of the local context is therefore essential as it may ultimately determine the success or failure of the business proposition or project intervention.

Supporting services

Supporting services form the third segment of the energy delivery model. These are the additional services which are not directly part of the market chain but can greatly influence the success rates of market chains through training and capacity building, microfinancing services, or other financial supporting services. Specifically, this category includes those services that the implementing actors in the energy chain have some control over, which differentiates this category from the enabling environment. The range of services that can potentially add value is huge and includes the supply of a range of products and services such as market information; financial and distribution services; quality assurance, including monitoring and accreditation; technical expertise and business advice; as well as support for product development and diversification. Table 2.1 details some of these supporting services as well as the actors who provide them.

Table 2.1 Supporting services and the actors who provide them

Supporting services	Actors			
	Government	Private sector	Donors	Civil society
Training/workshops	✓	✓		✓
Financing – grants/ subsidies	✓		✓	✓
Financing – loans, equity, debt	✓	✓	✓	
Technical support/services	✓	✓	✓	✓
Networking		✓	✓	✓
R&D	✓	✓		✓
Information exchanges	✓	✓		✓

Financial services

Financial services vary widely ranging from private-sector investments, in the form of debt and equity loans (e.g. private equity funds, venture capital funds, pension funds, and microfinance), to publicly supported instruments such as grants, low interest loans, guarantees, carbon credits or direct subsidies to reduce the final costs of the energy services (OECD/IEA, 2011). Some of the most important financing instruments, divided into both formal and informal, and financing sources for the three main energy delivery model distribution methods, are listed in Table 2.2.

Informal financing. Informal methods of sourcing funds, such as borrowing from friends and relatives, are very common in many developing countries. There are few requirements, and no proper accounting is required, however there are disadvantages such as limited fund availability, or high interest rates, especially from money lenders. One of the most common informal institutions is the rotating savings and credit association (ROSCA), whereby members contribute a prescribed amount of money regularly and then the

Table 2.2 Financing energy access

Scale of distribution system	Financing instruments	Financing sources
Off-grid (potentially mini-grid)	INFORMAL	
	Cash loans	Relatives and friends, money lenders, rotating savings and credit association
On-grid Mini-grid Off-grid	FORMAL	
	Grants/credit	Multilateral development organizations
	Debt financing	Bilateral development assistance
	Equity	Developing country governments
	Loans	Private sector
	Subsidies	Investment funds
	Guarantees	State-owned utilities
	Insurance	Microfinance institutions
	Microfinance	International banks
	Carbon finance	National development banks
	Feed in tariffs	Rural energy agencies/funds
		Foundations

total, or part, of the money raised is given to one of the members during each meeting until all the members have been serviced.

Formal financing. Financing is supplied through formal institutions, which are generally regulated by governments and lend money as a business, earning interest on their loans. These institutions range from microfinance organizations, which loan very small sums of money to individuals and groups, through to governments, which can provide grants and put in place subsidies, to large multinational institutions. These institutions then in turn loan very large sums of money for large-scale infrastructure projects, including grid extension and large renewable energy systems such as wind farms.

Private equity funds, venture capital funds, pension funds, and microfinance have played a key role in financing for energy access or even grid connections (OECD/IEA, 2011). A host of factors such as system location and scale, income profiles of potential customers, and available subsidies, dictate whether a project can attract private investors/operators (ARE, 2011). Banks also provide wide product range such as asset finance loans for energy business assets, emergency loans or top-up loans.

In the delivery of energy for the poor, subsidies are common for all types of grid or off-grid models. Subsidies may distort the market but also help to create markets by encouraging consumers to buy grid-supplied electricity or energy products. Universal pricing (the common practice of charging the same prices countrywide), also creates disincentives for electric utilities to serve rural markets, where costs tend to be higher (Barnes et al., 1994).

Cross-subsidies are another possible way of improving access to energy for the poor. Projects stimulating local economic development are often supported by the private sector, although public–private partnerships are an increasingly popular financing model, especially for larger investments.

Financing for or investment in energy projects that stimulate economic development at the local level are often provided by formal private sources, although governments can also be major contributors, especially if the energy installations are large. Financial institutions often have an inadequate understanding of how energy technologies or products work, especially at the rural level, and are thereby unable to structure their financing of products effectively. These knowledge gaps need to be filled through another supporting service – capacity building – as explained in the next section.

Capacity building

Capacity building is a wide-ranging term, which includes training, both technical and non-technical, knowledge management, and skill improvement for the actors involved. In this case it applies to the design, implementation, and ongoing management and maintenance of the energy delivery systems. The right mix of specific knowledge and skills is required at national, regional, and local levels to deliver energy services successfully. Capacity building involves increasing or improving the levels of competencies of institutions and individuals. While institutions and organizations focus on the appropriate structures and governance, individuals (including entrepreneurs) enhance skills to be more competent in the delivery of energy, directly or indirectly. For example, governments may benefit from capacity building through training and workshops which help participants learn how to design better regulatory frameworks, particularly pro-poor frameworks. Electricity utilities may benefit from learning more about efficient business models or utilization of IT systems, and ways to reach poorer customers. Entrepreneurs may need increasing knowledge and skills on the operation of businesses, including: using appropriate business models, accessing financing, and marketing. Communities may need training on management of the maintenance of energy systems, and users may need similar training to become aware of their options regarding energy services, issues of quality and certification, and their rights as consumers.

For financial institutions, developing a successful energy portfolio requires expert technical assistance as energy loans can often differ from other conventional loans (consider for example a loan for a solar system that uses the solar system as a guarantee asset). These energy products and services are new for many financial institutions, which often lack the required human resources and skill set to ensure they are developed and promoted effectively. This lack of resources then results in their low prioritization thus further exacerbating the problem. Technologies, and the earning and saving possibilities that these offer, together with knowledge of the customer base, are pivotal to effective operation. Training can be useful to employees of financial institutions in order to familiarize them with the workings of the market. It is critical that staff understand not only the energy products and services, but also the customers' cultural and economic preferences in order to decide on, and operate, energy portfolios. For example, loan officers who provide end-use consumer finance must be specifically trained in relevant energy technologies and what financial loans would be most appropriate for each one, including the repayment schedules, as they are the agents in direct contact with the client. Without the local market actors buying into the idea of a particular financial product, the initiative may be less successful. Also the absence of IT solutions and customer management systems may hinder the ability to serve clients effectively, and might require some form of support.

An important component of capacity building is technical support and advice. For example, for many government actors, technical support is often provided to ensure better institutional and regulatory development, or for specific project design and implementation. For entrepreneurs, especially small-scale producers, technical inputs are essential to enable them to produce and distribute quality equipment to the consumers. Capacity can be built upon the expertise and knowledge base that has been developed by various multilateral institutions and international agencies (AGECC, 2010). Once institutional capacity has been developed in a number of sectors – public, private, and civil society – the country is likely to take an upward trajectory and achieve its goals sustainably, including energy delivery, but this needs to be tackled in a systematic and active way, rather than just hoping it will happen. For successful energy delivery to the poor, capacity-building efforts will need to provide targeted and appropriate information to the various actors in order to strengthen both the enabling environment and the market chain.

Research and development

R&D is essential for finding innovative approaches to energy delivery, especially when the end users have very low purchasing power and the services need to operate in areas with poor infrastructure and in situations of market failure. Research is essential for helping provide accurate data to decision makers (e.g. energy companies, investors, and government departments) so they are able to make the necessary choices, including the development of policy measures and the design of intervening pathways. Moreover, research is a dynamic process where constant efforts are needed to keep results relevant for a constantly changing external environment. Research conducted at the local level also entails capacity building for technological developments that can emerge from local needs and practices, as well as the development of standards to ensure quality products are produced and meet the needs of end users (Modi et al., 2006).

An International Science Panel on Renewable Energies (ISPRE) 2009 report suggests that R&D can be targeted at different stages of the innovation chain, yielding benefits in the short term (up to five years), medium term (5–15 years) and in the longer term (15 years plus). While short-term R&D is mostly carried out by industry itself to improve well-established technologies, medium and long-term research is mainly needed to underpin long-term improvements and enable breakthroughs that will provide a decisive advantage in energy markets. This longer-term research is often significantly supported by the public sector and international donors (Modi et al., 2006). In energy access for the poor, donors – both bilateral and multilateral – also play a crucial role in providing funds for R&D to enhance technical as well as social and market development. The UN Global Environment Facility (UN-GEF) funding programme is a good example of a public-sector funding source supporting transfer of environmentally sound technologies to developing countries.

R&D for energy access is also particularly important in creating strong teams of knowledgeable and skilled people for the sector, an aspect that is widely recognized but not prioritized, especially in many developing countries. Research constitutes a strategically important position in the whole delivery chain and it is a crucial factor in determining the achievements that will be attained over time. In Africa, Lighting Africa, a joint International Finance Corporation (IFC) and World Bank programme, has been helping mobilize the private sector to build sustainable markets for solar photovoltaic (PV) technologies in Kenya and Ghana, providing safe, affordable, and modern off-grid lighting to some of Africa's largest off-grid populations. The programme had the goal of helping 2.5 m people have better lighting and improved access to energy by 2012. Market research was conducted, as well as technical research, to obtain a deeper understanding of issues associated with providing lighting to the customers at the base of the pyramid. The research was then shared with all relevant stakeholders. Thanks to their focus on R&D, the Lighting Africa programme has shown a high degree of success and its product certification is gaining the recognition of consumers. It is now being considered to be a benchmark for quality in other countries such as Uganda, Tanzania, and Senegal where it is hoped the programme will be replicated.

Knowledge management

Knowledge management is another area where supporting services can be invaluable, as improving the design of new energy delivery initiatives requires the taking into account of the knowledge and experience acquired in previous energy delivery initiatives. An important part of knowledge management is the capturing and sharing of knowledge, both expertise and information, within and between a range of organizations. This is a crucial driver for innovation since products and services are often greatly improved upon as experience is gained.

However, all too often, monitoring and evaluation (M&E) processes produce data which is not necessarily easy to use effectively. For example, lack of standardization of M&E practices makes it difficult to compare experiences and generalize lessons learnt, especially from larger samples. In addition, the format of evaluation reports, often a few pages of narrative, makes it impractical to even understand whether or not they might contain relevant insights. Moreover, the available evidence can be incomplete and/ or of poor quality, partly because organizations tend to highlight successes and cover up failures; successful solutions also constitute an asset against competition that organizations, especially private businesses, are reluctant to share.

Public databases and repositories can constitute invaluable resources for organizations that would not otherwise have the capacity to access that information. Examples of these are meteorological data, which has been made available by NASA, UN countries' census data, and so forth. The sharing of knowledge among practitioners is made possible through their participation in networks, communities of practice, or membership of associations such as the Household Energy Network (HEDON) and Practical Action's Practical Answers service which offers technical reports and a manned advisory service. Online platforms such as these can help enable exchange, dialogue, and partnerships which can lead to joint research or collaborative practices.

Categorizing energy delivery systems by distribution method

A convenient way of categorizing energy interventions, and therefore energy delivery models, used by most international institutions and organizations (e.g. the International Energy Agency, the UN, and Sustainable Energy for All, etc.), is through their methods of distribution – from on-grid, to mini-grid and off-grid. These methods are outlined in this section and then explored in much greater detail in Chapters 3, 4 and 5.

The three main distribution methods for energy delivery models are:

1. **On-grid:** centralized energy generation and distribution at a large scale (generally serving many thousands of people), usually referring to electricity and gas from national, regional and district distribution grids. On-grid system capacities are generally greater than 3 MW.
2. **Mini-grid:** localized, small or medium-scale power source (typically from 20 kW to 3 MW, although even smaller mini-grids are being trialled in countries such as India) supplying a local distribution grid connected to several domestic, business, and institutional customers in the locality.
3. **Off-grid:** stand-alone energy systems that serve individual domestic, business, and institutional customers.

Table 2.3 presents a concise view of the market chain, the supporting services, and the enabling environment, including social context, for the three scales of delivery. Each of these segments of the energy delivery model are expanded in much greater detail within Chapters 3, 4 and 5, to highlight the relevant critical issues related to each scale of energy systems. Each chapter focuses on the crucial issues which relate to the energy market chain processes (technology development and implementation, distribution of the conversion equipment and appliances, and the management and maintenance of the systems), the supporting services (financing of the interventions, capacity building, outsourcing, provision of complementary products), and the enabling environment elements (e.g. infrastructure, policies, regulations, social and cultural context), as well as the different roles and responsibility of the actors involved.

The use of such categorization is helpful in analysing the main ways that energy is delivered, and is of crucial importance in helping practitioners fully understand the main barriers and identify the potential solutions to provide a universal and sustainable supply of energy and help empower the marginalized to overcome poverty.

Table 2.3 Energy delivery models according to scale

Delivery model scale	Energy market chain				Supporting services		Enabling environment	
	Energy source	Technology	Management	Payment systems	Financial services	Actors	Policy	Socio-cultural context
On-grid	• Fossil or nuclear fuels • Large-scale renewables (mainly hydro but increasingly solar, wind, biomass, biofuels, and geothermal) • Hybrid systems (e.g. coal and biomass)	• Large combustion, nuclear and geothermal power stations • Large renewable energy-based power stations (large-scale bio-digesters, solar arrays, hydro dams or run-off river hydro systems, wind farms) • Energy predominantly reaches areas of society that are physically easy to reach (urban and peri-urban, and surrounding rural areas) due to infrastructure imbalance between urban and rural areas	• Generally initiated by government bodies and NGOs, often with backing from international organizations, and multilateral or bilateral donors • Implemented and managed by private sector with oversight from public sector	• Users pay for service either at regular intervals (e.g. monthly billing), or through prepaid metering system (purchase of credit and mobile payment systems, e.g. M-Pesa system in Kenya, etc.). • Maintenance paid for by the state/private utilities through billing • Service (especially for poor) is often subsidized by government or international donors • Typically requires a significant connection fee	• Government loans and grants (domestic finance including subsidies) • Possible corporate investment • National and international bank loans/guarantees	• Government- • Large-scale private sector (international and domestic) • Multilateral or bilateral donors / investors (e.g. World Bank) • International or domestic banks • Carbon brokers	• Government policies and regulatory frameworks influence different choices of fuels and technologies, including their investments • Regulatory bodies oversee infrastructure construction (particularly important for hydro) • International frameworks – especially for renewables • Carbon market frameworks and policies continue to be important • Specialized subsidy or tariff frameworks required for poor	• Energy can generally only be accessed by those who can afford to pay and who operate within established financial markets • Local expectations and preferences are often for grid expansion, even where the on-grid power is unreliable • Poor households continue to remain locked out of on-grid systems because of connection fees and high ongoing fees

(Continued)

Table 2.3 Energy delivery models according to scale (*Continued*)

Delivery model scale	Energy market chain				Supporting services			Enabling environment	
	Energy source	Technology	Management	Payment systems	Financial services	Actors	Policy	Socio-cultural context	
Mini-grid	• Fossil fuels • Small-scale renewables (including micro-hydro, solar, wind, biomass, biofuels, potentially geothermal) • Hybrid systems (e.g. diesel and pure plant oil generators such as multifunctional platforms)	• Generators (specifically designed for diesel, biodiesel or pure plant oil) • Micro plants (e.g. biodigesters, biomass boilers, gasifiers, solar, hydro dams or run-off river systems, and wind turbines) • Community mechanical power systems (e.g. grain mills and water pumping facilities)	• Often initiated by civil society (e.g. NGOs) and public sector • Management often by private sector with participation of community organizations, through NGO and public support, such as public–private partnerships	• Users pay for service either at regular intervals (e.g. monthly billing), or through prepaid metering system • Maintenance is paid for by the users, directly or indirectly • Systems often subsidized by NGOs and other donors to reduce payment levels • Sometimes include community-level technologies (e.g. community bio-digesters) with specialized maintenance and payment schemes	• Often involve some level of local authority/NGO/donor subsidy to pay for capital equipment • Impact/social investment • Venture capital (rare) • Corporate social responsibility (CSR) • Loan guarantees from some financial institutions • Difficult to scale up pilots without significant initial subsidising or appropriate financing loans	• NGOs • Community organization • Public-sector involvement (generally more local government departments) • Private-sector investment (limited but due to increase) • Financial institutions (mostly formal) • Carbon brokers • CSR	• Regulatory bodies, particularly for processed fuels • Renewable energy often prioritized due to donor/NGO support and high cost of transporting alternative fossil fuels • Policies such as tax and tariff reductions, and feed in tariffs are important contributors	• Often pilot schemes have supplied a range of energy services to isolated, energy-poor communities with strong local participation and involvement at all stages • Acceptance of new technologies can be an issue particularly if on-grid expansion is expected • Local willingness to maintain the service may be an issue if local skills are not present • Local ownership may be an issue	

Table 2.3 Energy delivery models according to scale (Continued)

Delivery model scale	Energy market chain				Supporting services		Enabling environment	
	Energy source	Technology	Management	Payment systems	Financial services	Actors	Policy	Socio-cultural context
Off-grid	• Fossil fuels (e.g. LPG cylinders and kerosene bottles) • Household/institutional scale renewables (including pico-hydro, solar, wind, biomass, and biofuels) • Hybrid fuel mixes (e.g. firewood, charcoal, and agricultural residue stoves)	• Household heating and cooking systems • Stoves (using charcoal, firewood, pellets, ethanol, or pure plant oil) • Solar water heaters and biomass boilers • Small generators (for diesel, biodiesel, and pure plant oil) • Solar home systems, solar lamps, solar fridges, single wind turbine, batteries • Household mechanical power systems (e.g. treadle pumps and individual grain milling systems)	• Private sector, particularly social entrepreneurs • Community-led initiatives • Household initiatives • NGO and donor initiatives	• Fuel and maintenance generally paid directly by users • Bioenergy mostly purchased by users through largely unregulated markets or obtained at no cost • Microfinance starting to be used for stoves and renewable energy systems (e.g. solar home system and solar lantern), as well as informal loans (e.g. SACCOs)	• Initial donor support to pilot new technologies and fuels until private sector can be built up • Subsidies required to ensure the needs of the poor are met • Carbon finance important • Impact/social investment • Venture capital (rare) • CSR (starting to increase)	• NGOs and local government act as facilitators • CBOs are important • Private-sector companies • Carbon brokers • Financial Institutions (formal and informal) • CSR	• Policy and national strategies for off-grid energy delivery, including decentralized energy production tax exemptions, subsidy policies, etc. • International frameworks for low-carbon/carbon markets • Regulatory bodies involved in supply of packaged fuel and technology products	• Packaged fuels often provide increased quality of energy (ease of use, efficiency and health effects) but due to their supply through markets require payment, which is often beyond the reach of the poor and marginalized • Innovative payment systems need to be developed to allow the poor to pay in very small increments matching their informal and irregular incomes

Note: MDG 1 – End poverty and hunger; MDG 2 – Universal education; MDG 3 – Gender equality; MDG 4 – Child health; MDG 5 – Maternal health; MDG 6 – Combat HIV/AIDS; MDG 7 – Environmental sustainability; MDG 8 – Global Partnership. All energy services can contribute to MDG 7 if low-carbon sources are used sustainably (Yadoo et al., 2011)

CHAPTER 3
On-grid energy delivery

After a brief outline of on-grid technology solutions and designs, this chapter will focus on how on-grid energy delivery systems in developing countries meet, and do not meet, the energy needs of the poor. On-grid delivery for the poor needs specific efforts in design and implementation, and appropriate financing arrangements. Most of the existing initiatives are not driven by profit alone and are innovative arrangements supported mostly with public or grant financing, buy-in from specialized financing institutions, and the localized provision of maintenance and empowerment services.

Keywords: distribution, energy delivery system, financing, on-grid, technology design

On-grid energy delivery refers to energy that is delivered through centralized power generation and distribution systems at a regional, national, or district scale, typically serving thousands of households, businesses, and public facilities. The International Energy Agency (IEA) estimates that more than 75 per cent of electricity consumed is distributed by on-grid systems in developing countries, although no definitive data currently exists. To support the direct delivery of energy through centralized grids, a parallel and intersecting market chain of energy products including metering and energy-powered appliances exists, allowing households and businesses to make use of the energy, providing the products are available and affordable. These products are generally supplied through private-sector companies with governments usually establishing an enabling environment, which ranges from stimulating the production of appliances to opening markets and helping them flourish (these aspects will be explored further in the 'Enabling environment' section of this chapter).

After a brief outline of on-grid technology solutions and designs, this chapter will focus on how on-grid energy delivery systems in developing countries meet, and do not meet, the energy needs of the poor. On-grid energy systems are often synonymously associated with electrification, but they have the potential to deliver a range of energy services, including heat and gas, although this almost exclusively only occurs in developed countries. In line with the theme of this book, this chapter will focus on the issues of on-grid delivery of electricity to the poor, both in urban and rural areas, and to a large extent will ignore the overall on-grid sectoral

http://dx.doi.org/10.3362/9781780447612.003

developments. As on-grid delivery for the poor needs specific efforts in design and implementation, as well as appropriate financing arrangements, examples are limited and most of the existing initiatives show that they are not driven by purely profitability goals alone. The arrangements are innovative, supported mostly with public or grant financing, buy-in from specialized financing institutions, and the localized provision of maintenance and empowerment services. Along the market chain, some of the electricity utilities partner with standardized product and equipment suppliers linking consumers through end-user financing support. This allows consumers to repay the electricity connection and appliances over a period of time through their monthly billings, to overcome the initial high costs. This requires an equal level of buy-in from the end consumers and the communities for the initiatives to be successful. Figure 3.1 provides an example of market mapping of Codensa, a Colombian utility which has developed a specific on-grid electricity supply model for the poor.

Market chain

Energy sources and technologies

Energy is typically delivered in developed countries through national, regional or district on-grid systems in the form of electricity, gas, and occasionally heating, and supplied through district distribution grids. In most countries, large-scale thermal power plants burning fossil fuels (coal, gas, and oil) dominate the electricity production sector. This method of energy production is starting to be complemented with co-firing in some developed countries, using biomass fuels, including wood and agricultural and/or municipal waste to meet renewable energy or energy security targets. By the end of 2011, around 45 GW of thermal capacity were being co-fired with biomass to some extent in Europe. In North America, around 10 GW of capacity is being co-fired with biomass (IEA Bioenergy, 2009). In developing countries, very little biomass co-firing has been installed, but as the technologies become more widespread and viable, and fossil-fuel prices increase, this is likely to change, particularly in countries where biomass resources are abundant. Nuclear power is also extensively exploited in many developed and some middle-income countries, including China, India, and South Africa. Similarly, on-grid district heating has generally only been installed in cold, northern developed countries, as outlined in Box 3.1, and is increasingly being delivered as part of combined heat and power systems in an attempt to increase the efficiency of energy production and delivery. Again, this could be replicated in some developing countries.

Enabling environment

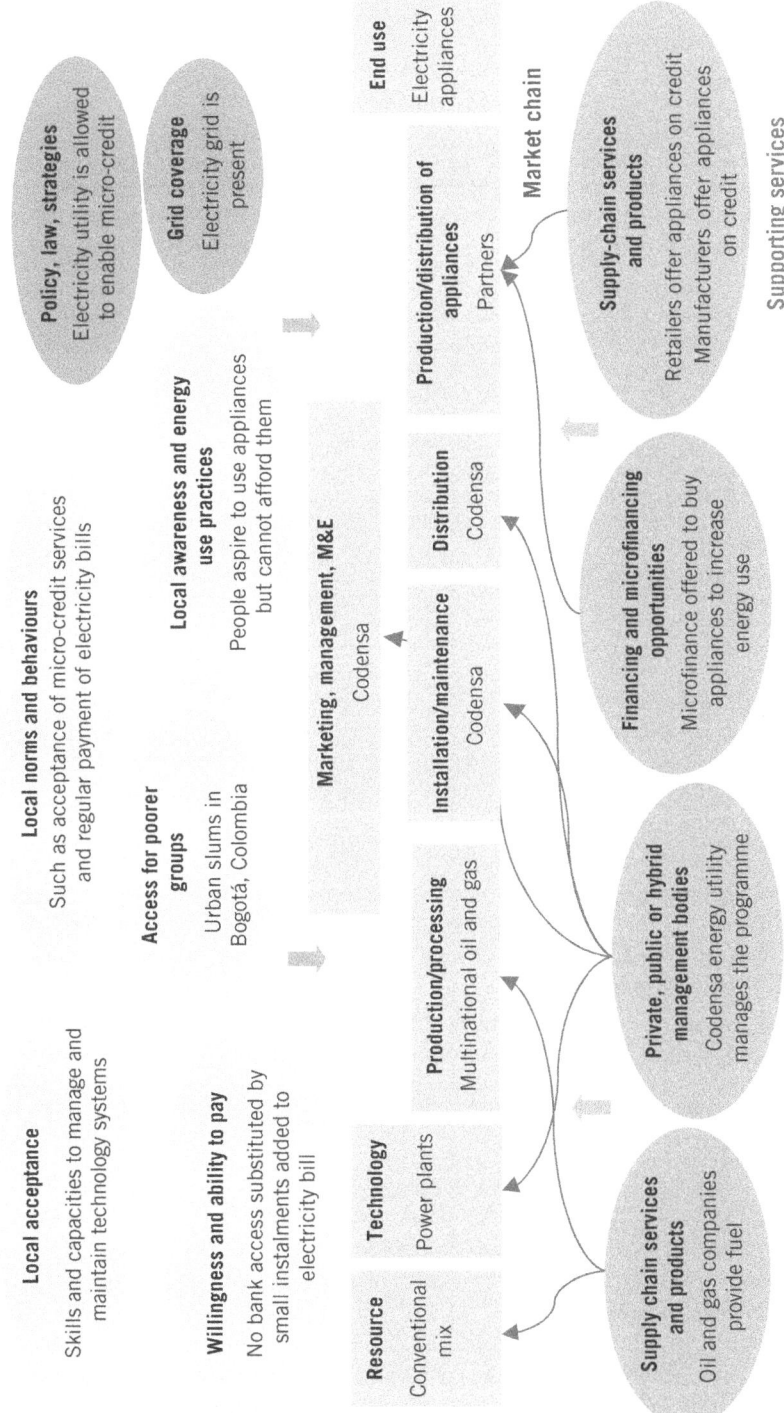

Local acceptance

Skills and capacities to manage and maintain technology systems

Willingness and ability to pay

No bank access substituted by small installments added to electricity bill

Local norms and behaviours

Such as acceptance of micro-credit services and regular payment of electricity bills

Access for poorer groups

Urban slums in Bogotá, Colombia

Local awareness and energy use practices

People aspire to use appliances but cannot afford them

Policy, law, strategies

Electricity utility is allowed to enable micro-credit

Grid coverage

Electricity grid is present

Resource	Technology	Production/processing	Installation/maintenance	Distribution	Production/distribution of appliances	End use
Conventional mix	Power plants	Multinational oil and gas	Codensa	Codensa	Partners	Electricity appliances

Marketing, management, M&E
Codensa

Market chain

Supply chain services and products

Oil and gas companies provide fuel

Private, public or hybrid management bodies

Codensa energy utility manages the programme

Financing and microfinancing opportunities

Microfinance offered to buy appliances to increase energy use

Supply-chain services and products

Retailers offer appliances on credit
Manufacturers offer appliances on credit

Supporting services

Figure 3.1 Market map of Codensa on-grid energy delivery

Box 3.1 Examples of district heating

District heating systems were common in the former Soviet Union and account today for more than 70% of all heat distributed in Russia as the nation's single largest product market, worth an estimated US$33 bn in annual sales (Johansson et al., 2010). Other countries also started to adopt district heating decades ago, and Denmark, Germany, and Sweden, for example, are now the largest users of municipal solid waste for this purpose.

Some countries, particularly in Scandinavia, show a significant penetration of district heating of over 50% of the heat market. District heating and Cooling (DHC) only currently constitutes a small fraction of the total heat market of the EU, but the un-tapped potential is large, varying in each country depending on past national policies. In the US and in other countries where much energy is used in cooling, use of district cooling has grown significantly.

While DHC has only been of importance in northern latitude developed countries, typically Europe and North America, it is unlikely that it will be of major significance in developing countries, as they are predominantly quite warm; however, increasingly, many parts of the world are implementing DHC for cooling, either through distribution of chilled water or by using the district heating network to deliver heat for heat-driven chillers. This may become more common in some developing countries in the future, particularly large urban centres.

Source: OECD/IEA, 2010

The dependency on fossil fuels is, however, changing as countries move towards low-carbon development, and as renewable energy technologies become increasingly cost-competitive.

Renewable energy

Large-scale renewable energy production systems are being deployed in increasing capacity to supply on-grid systems, with some countries now producing a significant proportion of their national energy usage, or even exceeding internal demand and exporting the excess to neighbouring countries. Examples of this include Laos and Uruguay. In Laos, 99 per cent of the total installed electricity capacity is from hydropower resources (Reegle, 2006).

Hydropower

This is the largest source of renewable power generation worldwide and yet it has been estimated that less than one-quarter of the world's technical hydropower potential is currently in operation (IRENA, 2012a). In 2009/2010 the global installed hydropower capacity was estimated to be approximately 940 GW (excluding pumped storage hydropower), provided by 11,000 hydropower plants in 150 countries. This accounted for 16.5 per cent of global electricity production and 85 per cent of total renewable electricity generation, and supplied more than 1 billion people with power (REN21, 2011). The world leaders in hydropower are China, Brazil, Canada, the US, and Russia. Asia accounts for the largest share of global installed hydropower capacity, followed by Europe, then North and South America, then finally, Africa (WEC, 2010 and IHA, 2011). Hydropower from the Itaipu dam in Brazil, for example, generates over a quarter of the country's electricity supply, along with 100 per cent of Paraguay's (OECD/IEA, 2012).

In Africa, there are a number of countries that produce a significant proportion of their grid-based electricity from hydro, such as the Democratic Republic of Congo (DRC) and Ethiopia. As a whole, hydropower accounts for approximately 32 per cent of current capacity, but this capacity is estimated to be 3–7 per cent of the technical potential of the African continent (IRENA, 2011). In Ethiopia, hydropower energy is the focus of government plans for energy supply, with up to 90 per cent of their electrical energy coming from this source (Heimann et al., 2008). Although the focus has historically been on large-scale hydropower plants, their output is being affected by changing rainfall patterns (often linked to an increasingly changing climate) and in some cases lack of maintenance, which has caused many plants to fall into disuse, particularly in East Africa.

Large-scale wind farms

Both onshore and offshore wind farms are also starting to contribute to the on-grid energy mix. The global wind power market continues to grow, with 238 GW installed by 2011, and the main drivers being China and India, which together account for just over 50 per cent of the global market. Starting from 2010, the majority of wind power installations were outside of the Organisation for Economic Co-operation and Development (OECD), and according to the Global Wind Energy Council (GWEC), this trend is likely to continue in the near future, with Brazil and Mexico expected to be the major wind energy growth markets in the coming years, as well as countries such as South Africa and Namibia (GWEC, n.d.).

Solar technology

This technology is increasingly playing a role in on-grid energy systems, with solar photovoltaic (PV) being one of the fastest growing renewable energy technologies. Driven by attractive policy incentives (e.g. feed-in tariffs and tax breaks) in the last few decades grid-connected installations have become the largest sector for new PV installations. The global installed PV capacity grew at a rate of 44 per cent in 2011 to 67.4 GW, with Europe accounting for around three-quarters of all new capacity added in 2011, and six countries adding more than 1 GW in 2011 alone (Italy, Germany, China, the USA, Japan, and France). This rapid expansion in capacity has led to significant cost reductions, but despite the rapid growth of the PV market, less than 0.2 per cent of global electricity production is currently generated by PV. However, given the pace of recent developments in the PV sector, predictions by the IEA's PV roadmap (based on 17 per cent market growth rate) indicate that installed capacity of solar PV for on-grid energy systems will reach at least 200 GW by 2020.

Another solar technology on the rise in supplying on-grid energy systems is concentrated solar power (CSP), with the global installed capacity in 2012 being around 1.9 GW but with dozens of plants under construction. The two leading concentrated solar power (CSP) markets are Spain and the US with 90 per cent of the global share. The markets are driven though tax incentives and renewable portfolio standards in the US, and feed-in tariffs in Spain. Algeria, Egypt, and Morocco have built, or are building, CSP plants which are integrated with their natural gas combined-cycle plants. Australia, China, India, Iran, Israel, Italy, Jordan, Mexico, United Arab Emirates (UAE), and South Africa are developing large-scale on-grid CSP plants, which will be coming on-line in the coming years (IRENA, 2012b). Several developing countries have been investigating the potential of CSP for supplying their on-grid energy systems, particularly in sub-Saharan Africa, and as the market matures and costs start to come down, it is likely that investment in on-grid CSP systems will start to take place.

Geothermal and bioenergy

This is another on-grid energy resource that has great potential in a number of developing countries, although most of the installed capacity is in developed or emerging middle-income economies. The total installed geothermal capacity in 2010 was nearly 11 GW worldwide, with the US contributing the largest share of capacity, followed by Mexico, Italy, New Zealand, Iceland, Japan, the Philippines, and Indonesia (IEA Geothermal, 2012). In East Africa, Kenya has started the development of a number of geothermal systems, which are expected to contribute significantly to their total energy production, predicted to soon overtake hydro systems that are starting to become increasingly unreliable due to changing rainfall patterns.

In 2010, the global installed capacity of biomass power generation plants was between 54 and 62 GW, providing around 1.4 to 1.5 per cent of global electricity production (REN21, 2011). Europe, North America, and South America currently account for around 85 per cent of total installed capacity globally; in Europe 61 per cent of installed solid biomass capacity (excluding wood chips) is in England, Scotland, and Sweden; and it is expected to continue to grow in the near future as developed countries try to reach their renewable energy production targets. Bioenergy is usually not viewed as a viable renewable energy sources in many developing countries, as it is associated with traditional energy usage in rural areas, and is not seen to be a modern energy resource. Despite the large biomass resources in developing and emerging economies, the relative contribution of biomass towards on-grid energy production is negligible, even though it has great potential for reducing energy poverty. The political discourse around the international trading of bioenergy, in particular biofuels, is ongoing and likely to continue into the future. It includes the controversies of land grabbing, biodiversity loss, and competition with food production.

Bioenergy has great potential for reducing energy poverty in a number of developing countries, and local utilization should be prioritized above trading it internationally. In Latin America, Brazil is the largest producer of biomass electricity as a result of the extensive use of *bagasse* for co-generation in the sugar and ethanol industry, and several countries in Africa are trying to replicate this success. As bioenergy technology systems mature and models of effectively managing forest and agricultural resources are developed, it is expected that government support for using bioenergy for supplying on-grid energy systems will grow in many developing countries, to contribute significantly to their national energy portfolios (Macqueen 2011; Cotula et. al., 2011; Practical Action Consulting, 2012).

Small-scale on-grid connections

In most emerging and middle-income countries, a range of small-scale energy technology systems are increasingly being installed to feed directly into the grid. In particular, small (under 10 MW) and micro (under 100 kW) hydro, solar, and wind power systems, and small-scale gas systems (typically 1–2 MW community gas-to-power plants) have started to be promoted through innovative subsidy schemes, such as feed-in tariffs. On-grid connections are important as they have the potential to deliver energy to the nearby populations where the electricity is being produced. It is believed that they will start emerging as a significant opportunity in developing countries in the next decades but within the context of this book they are not taken into consideration as there are currently very few installed examples.

Ownership, management, and maintenance

Government bodies typically initiate on-grid energy delivery, mostly through their ministries of energy (or equivalent). Where a country's national budget is not able to meet the expense of these large infrastructure systems, financial backing is obtained from international organizations; multilateral or bilateral funders, and regional development banks (e.g. the African Development Bank – AfDB, Asian Development Bank – ADB, Islamic Development Bank – IDB, and the Inter-American Development Bank – IADB). Most on-grid systems consist of a national-scale electricity grid supplied through power cables, initially in urban areas gradually extending into rural areas.

In many developing countries, governments usually remain the owners, managers and distributors of the national and regional grids, but may contract out various parts of the supply chain – construction, maintenance, billing – either partially or completely to private-sector organizations, through different models, as detailed in the following section. The cost of grid extension into rural areas is extremely high, and is often far from being cost-effective, due to the relatively low population density and low level of economic activities. The World Bank's worst-case scenario estimates that grid extension could cost up to $19,000 per km in some countries, providing a significant economic barrier to expansion (ESMAP, 2000). An indirect implication is that where the delivery services are managed purely by private companies, the incentive to expand into rural areas is frequently very low, both due to installation costs and the uncertainty of cost-recovery through customer payments. As a result, the poorest and most isolated are generally the least served by on-grid systems. Additionally, on-grid energy systems in more rural and remote areas tend to suffer from significant levels of inefficiency, poor maintenance, and substantial losses, particularly in many parts of sub-Saharan Africa, as detailed in Box 3.2. This leads to increased prices, as well as power disruptions and inequality in distribution, often with the poor and marginalized being the worst hit. Power companies are able to control where the power is directed and often shut down power in the poorer areas first.

Box 3.2 Inefficiencies in on-grid electricity supply

The efficiency of on-grid electrification systems in developing countries is often quite low due to high transmission losses (in this context, losses refer to the amount of electricity supplied through the transmission and distribution grids but not paid for by the end users). Total losses consist of technical and non-technical components. Technical losses occur naturally and consist mainly of power dissipation in electricity system components such as transmission and distribution lines, transformers, and measurement systems. Non-technical losses are caused by actions external to the power system and consist primarily of electricity theft, non-payment by customers, and errors in accounting and record-keeping (World Bank, 2008). In many developing countries, theft and poor cost recovery through inefficient billing mechanisms are the most severe, and national electrification utilities are continually developing strategies to try to reduce these non-technical losses.

In sub-Saharan Africa approximately 50% of electricity generated is paid for, due to a combination of low percentages of electricity injected into distribution networks being billed, and low rates of collection of the billed amounts. The variation in performance is enormous, with the highest inefficiencies in Nigeria, where the national utility only captures 25% of the revenues owed to them. Some recent studies have shown that hidden costs of distribution losses in sub-Saharan Africa are usually more than 0.5% of gross domestic product (GDP), and may be as large as 1.2% of GDP in some countries.

As mentioned, such inefficiencies can be reduced through proper management systems as demonstrated by the state-owned and operated utilities of Botswana and South Africa. Botswana Power Corporation has long provided a reliable and high-quality service, expanding the network in both urban and rural areas, covering its operational costs and posing no burden on the government budget. It has also reduced system losses to 10% and earned a decent return on assets. Electricity losses in Botswana are lower than in South Africa, where the power sector is operated by Eskom (one of the largest utilities in the world) with about 15% total losses.

Source: World Bank, 2008

The management arrangements for on-grid energy delivery vary widely, ranging from those fully under the control of the state, to those that are leased to external companies. These are discussed in the following sections. As indicated previously, on-grid models are mostly public-sector driven and managed in the developing countries, but innovative delivery mechanisms have stemmed from the involvement of the private sector and civil society, for example public-private partnerships.

Public–private partnerships management model

A common management model, especially for the poor or underserved in developing countries, is the public–private partnerships (PPPs) model, whereby the provision of energy (predominantly for electricity, but potentially for other types of energy services such as heat) is contracted out by the public-sector authority to a private company, who then assumes the management and operation of the system. PPPs allow for 'the efficiency, delivery capacity and resources of the private sector to support government objectives in expanding access and delivering clean, reliable and affordable energy services' (WBCSD, 2012: 14). This is a well-established and proven model in many parts of the world, and is increasingly being replicated to target poor, and energy-deprived, communities, as in the case outlined in Box 3.3. This box demonstrates how member companies of the Business Council for Sustainable Development in Argentina have significantly expanded gas and electricity connections within poor areas of Buenos Aires.

An example of another successful PPP comes from Morocco. In 1997, Lyonnaise des Eaux de Casablanca (LYDEC – a subsidiary of GDF SUEZ) formed a partnership with local authorities and the National Initiative of Human Development in the Greater Casablanca region of Morocco. The partnership took charge of a management contract to provide electricity, water, and sanitation facilities. One of the key reasons for this partnership was to address the high number of illegal and dangerous electricity connections and poor facilities in the shantytown areas of Casablanca. LYDEC employed several innovations to try to ensure that 75 per cent of the households in the targeted areas had a safe and secure electricity connection within a five-year implementation period. A community connection model was developed where LYDEC supplied electricity to one community representative who subsequently sold the supply to 20 households in their block. This resulted in a reduction of the capital and operational costs required for providing the services. Customer contributions to support the connection costs were spread over three years to improve affordability and the reduction in costs was achieved through the use of trained local electrical contractors and the adaptation of equipment standards. In addition, a partnership model for project design and implementation to ensure buy-in from local communities was developed, and an open dialogue between partners on performance, decisions, and accomplishments took place.

Box 3.3 Public–private partnerships in Buenos Aires, Argentina

The Business Council for Sustainable Development in Argentina (or Consejo Empresario Argentino para el Desarrollo Sostenible – CEADS) was set up in 1992, bringing together the chairpersons of 16 prominent Argentine companies to stimulate private-sector involvement and investment in conjunction with public-sector delivery targets, including developing business solutions to energy access challenges in the country. Among the initiatives supported by CEADS are the two following examples that show how a partnership between various private-sector actors and government authorities can be effective in delivering energy to marginalized urban populations.

The first case is Empresa Distribuidora y Comercializadora Norte S.A. (EDENOR), an electricity distribution company, which works with 2.5 million customers in northern Buenos Aires. It has developed a number of new initiatives for maintaining access to electricity for poor families, and for reducing illegal connections. These include: converting households to pre-paid meters; providing households with more direct control over their electricity use; and helping them avoid payment difficulties associated with standard 60-day billing. The company has complemented these measures with a household energy efficiency programme aimed at helping customers reduce their consumption. The overall impact of these measures has produced average cost savings of 37% for customers, without affecting quality of service.

The second case is Gas Natural Fenosa, a company that has developed an inclusive business model with the Fundacion Pro Vivienda Social to support extension of the natural gas network in low-income neighbourhoods on the outskirts of Buenos Aires. It involved the creation of a community-owned trust fund which provided a collective guarantee to pay for the extension of services, as well as supporting the network extension. The model also helped individuals to obtain access to microcredit and to pay for their individual connection costs. The project has resulted in the extension of on-grid services to 11,000 extra people in its first five years, and led to significant reductions in household energy expenditure (from 13.8% to 3% of income), demonstrating the benefits for households and providing additional revenue to support their investment in a network connection. Funding for the project came in the form of a $1.7 m loan from social investors (World Bank and Fondo de Capital Social – FONCAP), which was repaid through individual household billing. The project has resulted in the improvement of the living standards of over 3,000 families, through a reduction in respiratory illnesses and better energy access, and with average annual family savings of $14 on energy expenditure. It is expected to be expanded to reach a further 10,000 families by the end of 2013.

Source: Aron et al., 2009; WBCSD, 2012

Along with monthly profits of $270, the fact that a whole block was disconnected if one bill was unpaid ensured the community representatives in charge of each block had an important incentive to operate effectively; this resulted in 98 per cent of bills being paid on time over the five-year period. LYDEC was experiencing a $1.4 m annual loss due to fraud and illegal connections before the project, which was replaced by an overall profit of $400,000 by the end (WBCSD, 2012).

Experience has shown that successful management alone is insufficient to effectively deliver on-grid electricity; the right mix of supporting services, in the form of local ownership and public support, financing, and training from the utilities, is of equal importance for on-grid electricity services to be successful, particularly when targeting the poor.

Cooperative model

Cooperatives, owned and managed by community members, have also been involved in rural electrification, commonly under a public–private arrangement. Costa Rica, Bolivia, Bangladesh, India, and Nepal are some countries where cooperative models have been used to increase on-grid electrification to rural areas. Experience varies, with cooperative models either acting as stop gaps for reaching rural areas in the short term, or as part of mainstream electrification initiatives. In Costa Rica, a rural cooperative was supported by USAID with contributions from the National Bank of Costa Rica, as well as start-up funds from the cooperative itself, to cover the capital costs of on-grid electrification. The main impact was to lower the electrification costs which were previously provided by private run mini-grids from diesel generators. These generators were very expensive and reached far fewer households (ESMAP, 2005). By 2011, only a handful of rural cooperatives producing, distributing, and marketing power in rural areas were not a part of the two main state-owned enterprises – the Instituto Costarricense de Electricidad (ICE) and its subsidiary, Compañía Nacional de Fuerza y Luz (CNFL) (Reegle, 2011) – showing the success of the initiative.

In Bangladesh, cooperatives are part of the rural electrification programme to publicly support electricity distribution businesses (Barnes et al., 2010). A Rural Electrification Board (REB) was established in 1977 as a semi-autonomous government agency with 70 operating rural electric cooperatives called Palli Bidyuit Samity (PBS) covering more than 90 per cent of the area for rural electrification (Reegle, 2012).

In Nepal, community-based cooperatives invest 10 per cent of the cost of the systems, complemented by a 90 per cent investment from the government, managed through the Nepal Electricity Authority (NEA) which has developed into a remarkable and popular mechanism for on-grid rural electrification. The scheme started out with a contribution of 50 per cent from the Government of Nepal, with the remaining 50 per cent required to come from the community itself. However the government contribution had to be gradually increased in order to reduce the financial burden on the community and to meet their basic infrastructure needs. Due to this flexibility in approach, the pace of electricity distribution in Nepal has risen significantly over the past six years. In addition, the introduction of the Nepal Electricity Authority Community Electricity Distribution Bye Laws, 2060 (which allow community participation in rural electrification activities), have helped the NEA cooperative electrification model bring in 230 Community Rural Electric Entities (CREEs) to electrify 211,942 households since 2005, with a further 107,000 households due to be electrified in the near future in 46 districts of Nepal (Prasad, 2013). The programme has been successful due to the range of capacity-building initiatives, and the promotion of productive uses of energy in order to contribute to the economic development of the rural regions of Nepal.

Franchising

An alternative on-grid energy delivery management system is that of franchising. This model of electricity distribution involves a public or private grid-distribution company franchising the operations of a particular area to a local enterprise. The enterprise initially receives external support for the start-up costs, but then covers the ongoing costs through the collection of the local tariffs paid by the end users. The delivery process of a franchise can vary, but is often focused on the operations that take place at the local level, such as maintenance and billing. Franchising models are not very common and as the case from India highlights in Box 3.4, may not be entirely sustainable or scalable without a strong commitment from the local utility companies and political parties, and concentrated local awareness-raising.

Box 3.4 Franchisee model in India: experience of S³IDF and ASCI

Universal access to electricity infrastructure is important for equitable development and continues to be an objective for the Government of India. Providing grid electricity, especially in rural areas, has been challenging, particularly a good quality and reliable supply. While the necessary legislative framework has been established and the government has committed finance for these innovations, demonstrated models of electricity distribution franchising do not readily exist in India. An innovative strategy to overcome this challenge is the concept of a local electricity distribution franchise piloted by the Small Scale Sustainable Development Fund (S³IDF) and the Administrative Staff College India (ASCI) in the *Ranga Reddi* district of *Andhra Pradesh* in India. Initially funded by Global Village Energy Partnership (GVEP) International, the project aimed to fill this gap and lay the foundations for replicable, scalable, and pro-poor models of franchising through a pilot project. Its central focus was an innovative institutional intervention that would empower local communities by creating an infrastructure management and service-delivery entity. This would then enable local franchises to bundle and facilitate other service delivery, such as financial and technical services, and energy conservation.

After several trials and negotiations with Andhra Pradesh Central Power Distribution Company Limited (APCPDCL), an 11 kilovolt (kV) feeder was established in the village of *Cherlapatelguda*, serving 630 households, 8 commercial connections, and 370 agricultural pumpsets. All APCPDCL staff who looked after the distribution line were withdrawn, and a local person appointed as legal franchisee with ASCI and S³IDF as mentors. Two senior officials who had retired from service were appointed for accounts, collection, operation, and maintenance. The initial phase was successful as it resulted in new connections and some results were:

- **Supply improvements:** Sub-station, feeder, village maps were prepared, 11 kV lines were surveyed, and poles numbered and coded. The defective LT lines and digital transfer ratiometers (DTRs) have been mostly rectified with the failure rate down to less than 1% out of 1,073 DTRs; transformers (including agricultural) have been provided with earthing; all low level distribution wires were properly tightened with additional poles; 100% metering was ensured for household consumers; and monthly/bi-monthly were bills prepared and handed to the customers.
- **Social mobilization and management:** Energy committees were set up in each village, and periodical meetings conducted. Vegetation management under the overhead lines and around DTRs was regularly maintained. In addition, 32 *vidyuth* (electricity) part-time volunteers were selected from the village, and consumer emergency calls were attended within hours, on the same day. However, while billing of the metered domestic connection had increased, collection was a challenge due to state-level political problems.

APCPDCL evaluated the work in 2008 and requested continuation as well as expansion to a larger area. However, the local franchisee was dropped as collection of fees was in great arrears due to political intervention. S³IDF took over as the new franchisee for technical maintenance of the distribution system, metering, and billing, while collection was carried out by the officials of the Andhra Pradesh Central Power Distribution Company Limited (DISCOM). A two-year contract was signed to cover 32 villages in *Yacharam Mandal*. However, despite the achievements, and while the implementation was highly appreciated by the senior officials, in March 2013 the junior executives of DISCOM and the workers opposed the extension of the franchise system as it was seen as a step towards total privatization and a loss of employment for themselves. Support also came from political parties who were opposed to any increase in the tariff and to the collection of a surcharge due to fuel price increases. Thus, from June 2013, all responsibility were handed back to APCPDCL, with S³IDF and ASCI continuing to work independently as NGOs for the popularization and deployment of renewable energy home lighting systems, especially for the poor.

This case study shows that the success of a franchisee model requires considerable external up-front support from the electricity distribution company (either public or private), full assistance towards building local capacity of an electricity distribution system, and support from all political groups. Public–private partnership models can become highly sensitive when issues of scaling up take place. Lastly, comprehensive consultations with electricity consumers on all aspects of the electricity environment and climate change consequences need to take place before any franchisee system is attempted.

Source: GVEP International, 2008; S³IDF and ASCI, Direct communication, March 2013

Payments systems

The cost of maintenance of on-grid electrification systems (the fuel-processing plants, generators, hydro dams, power line networks, and household connections), is typically recovered through the payments for the end-use energy services from households, public services, and businesses. In some countries, public utilities recover the costs directly, or contract the payment collection out to private-sector companies, either directly or as government-led concessions. End users pay either through regular payments, such as monthly billing, or less commonly through pre-paid metering systems. Such formal payment systems are generally acceptable for the wealthier households and businesses that have regular incomes, but are often not suited to poorer households, and more informal businesses, due to their significantly lower and more inconsistent incomes. Innovative payment schemes have been developed to try to address these payment barriers, such as the 'pay as you use' payment system which can be managed through mobile phones, based on Kenya's M-Pesa mobile payment scheme, as outlined in Box 3.5.

When an on-grid energy delivery system is to be extended to a new area, a key consideration is to ensure that the target consumers are able to afford the costs associated, including any connection charges, as well as the ongoing payment schedules. Gauging the local population's willingness, and ability, to pay for a service at a certain price is essential in planning the financial feasibility of a project. Often, if the delivered energy does not provide an economic benefit to the end users they will not be able to pay for the tariffs (NRECA, 2010).

Box 3.5 Innovative payment methods in Kenya

The M-Pesa scheme is a mobile phone-based money transfer facility, developed by Safaricom (an affiliate of Vodafone), and originally funded by the UK's Department for International Development (DFID). The system allows users to deposit, withdraw, and transfer money, pay bills and purchase credit through their mobile phones. The concept has proved to be extremely successful, resulting in nearly 10 million Kenyan adults gaining access to financial services they previously did not have.

A key benefit has been the ease with which microfinance institutions (MFIs) have been able to offer competitive loans to finance income-generating activities, especially energy delivery programmes. This has subsequently allowed many people the opportunity to effectively finance and manage their energy supplies. The scheme has spawned over 70 similar initiatives worldwide since its introduction in 2007.

Although the system was originally developed to try to help poor energy users access electricity supplied through on-grid energy delivery systems, it is also starting to be used within mini-grid and off-grid energy systems, such as M-Kopa, as explored in Chapters 4 and 5.

Source: Graham, 2010

An alternative method for helping the very poor access energy services through on-grid energy delivery models is the use of varying tariffs, with the rates increasing with the amount of electricity being used, and the lowest rates being applied to the first amount of electricity supplied. This helps poor households and small, often informal, companies access a limited amount of energy at a very affordable price while not being a large financial drain on the distribution company. The costs of the service are covered by the higher charges made to the larger energy users (wealthier households and businesses) who have higher incomes. An example of this method has been implemented by the South African government who have provided a 20-amp electrical connection to the poor, free of charge. In addition, most municipalities offer a free monthly allocation of the first 50 kWh of electricity used, where economically feasible. This model has led to extensive debates regarding its effectiveness in reaching the very poor, who often have not yet been connected to the grid or live in less wealthy municipalities that cannot afford to offer the roll-out of the free basic electricity. In addition, in order to avoid high administration costs incurred in verifying the eligibility for free basic supply, some municipalities simply apply the subsidy to all domestic consumers, including richer ones who happen to reside in those areas. Despite these limitations, South Africa's model and its tariff regime, which is one of the lowest electricity tariffs in the world, has been one of the most successful models for helping the poor access on-grid energy (Niez, 2010).

In countries such as Sri Lanka and China, a high percentage of the population are able to afford energy services from on-grid energy delivery systems due to their relatively high incomes (Van der Vleuten et al., 2007). However, 'middle-poor' households may be able to afford the access to a limited supply of electricity, but are often unable to afford the electrical appliances that then allow them to improve their standards of living or increase their incomes. In an urban slum in Colombia, the utility company,

Codensa, tried to address this issue by providing access to credit at an affordable interest rate to buy electrical appliances which was then charged as part of the monthly electricity bill, at the same time fostering the demand for electricity, as outlined in Box 3.6.

Box 3.6 Slum electricity connections by Codensa, Colombia

Codensa is a Colombian subsidiary of Edensa (an Argentine energy utility company), which found that it was unable to increase its customer base in Bogotá due to government restrictions on its market share, so instead developed a model to try to increase the energy consumption of its existing clients, many of whom were very poor. The company developed a complementary business to its delivery of on-grid electricity in urban slums, by offering microcredit to households to purchase electrical appliances in order to boost their electricity demand. The slum populations in the city had traditionally been unable to pay for the up-front costs of many electrical appliances, thereby keeping electricity demand and revenues low.

Codensa found that 66% of its customers had no bank access, and hence were unable to secure funding for purchase of appliances to improve their living standards. By forming a partnership with 18 retailers and 120 appliance manufacturers, they were able to offer credit to customers that could be repaid through monthly billing. Over 550,000 people from the lowest income strata in the region are now able to access credit facilities to purchase a wide range of electrical appliances. This has resulted in an increase in Codensa's revenues by approximately 40%, without an increase in their customer base. The advantage of the utility offering credit was that it could collect the repayments at a marginal management cost by including them as part of its standard billing.

It should be noted that before introducing such a scheme it is important to understand the incomes of the energy users, to ensure that they are able to afford to pay for the energy appliances, and do not get themselves into debt which they are unable to repay.

In Figure 3.1, the authors have provided a market map of Codensa showing clearly how the market chain is supported by the enabling environment and support services.

Source: Aron et al., 2009; Rufin and Márquez, 2011

Supporting services

Services that support the construction or use of on-grid systems are typically provided by the private sector. A thriving ecosystem of enterprises is required to build and maintain the distribution grid. Healthy and functioning on-grid energy systems need sufficient skilled labour both at the technical and organizational level, to effectively deliver energy all year round, and provide critical support services. This is particularly important for ensuring that on-grid energy systems are cost-effective and not highly resource-wasteful, which is often the case in developing countries. The development of innovative supporting services for on-grid energy systems is of particular importance for ensuring the poor can access energy services, as they are typically more difficult to reach. As well as the more traditional supporting services, an aspect not to be overlooked is the ability to supply appliances that enable customers to make use of the energy provided. Again this service is usually entirely the task of the private sector.

Many of the case studies presented in the 'Market chain' section of this chapter are successful because of the level of innovative financial services available, coupled with building the capacity of stakeholders involved along the delivery chain. The following sections will explain some of these in detail.

Financial services

On-grid financing is extensive and can involve direct investments to credit lines, guarantees, and in the case of many developing countries, grants. Often the financing of on-grid energy systems is achieved through a combination of debt, equity, and other sources of finance including grants, particularly for technical assistance, to achieve commercial success. While governments play a central role through public financing, the role of private-sector financing is also critical for driving the energy markets, as it is embedded in the market chain. In the delivery of on-grid services, customers (industry, entrepreneurs, and households) play an important role in creating a demand, and importantly, pay for the cost of the on-grid connection, or the appliances that are used. Developers of on-grid projects need to invest capital up-front, and are only willing to do so if they believe a return on investment is likely, especially as the size and scale of the investment rises.

The vast majority of financing for energy delivery has been directed towards large-scale on-grid electricity infrastructure. While this has been relatively successful in allowing people in developing countries to access energy it has generally not benefited poor households in urban areas, and most households in rural areas (OECD/IEA, 2011). As urban areas in many developing countries rapidly expand, there is a corresponding growth of poor people living in these areas (University of Michigan, 2002). The need to create innovative ways of delivering on-grid electricity has been initiated in a few countries, some of which are mentioned in this section; however, the cases are few, and innovative delivery systems for the poor are limited. There is a clear reliance on grants or revolving funds to initiate any financing mechanisms to help the poor or those on lower incomes benefit from on-grids. It is also important to look at the role of subsidies in the delivery of on-grid electricity. Subsidies are initiated as a policy measure but play an important role in the financing mix and are a critical supporting service for delivering energy access to the very poor.

Subsidies

A characteristic of many government-led on-grid energy delivery models is the use of subsidies. The IEA's latest estimates indicated that subsidies relating to fossil-fuel consumption worldwide amounted to $409 bn in 2010, with subsidies to oil products representing almost half of this total (OECD/IEA, 2011). The largest subsidies are generally offered in developing countries in an attempt to shield their populations from high fuel prices, which can limit development and growth. However, in reality, subsidies are often an extremely inefficient method of helping the poor access energy as they often end up benefiting the higher income groups, the fossil-fuel companies, and the technology producers. According to the *World Energy Outlook*, only eight per cent of the $409 bn spent on fossil-fuel subsidies in 2010 went to the poorest 20 per cent of the population (OECD/IEA, 2011) and in fact the International Monetary Fund (IMF) has established that, on average, over 80 per cent of fuel subsidies benefit the wealthiest 60 per cent of the population of developing countries (Arze del Granado et al., 2010).

This can be explained by ineffective targeting mechanisms, inadequate distribution schemes, and the fact that higher-income households consume more petroleum-based fuels (Global Subsidies Initiative, 2011). Moreover, direct subsidies focus on tariff reductions draining resources that could otherwise cover the cost of on-grid electrical connections, thereby actually becoming an obstacle to on-grid energy delivery for the poor. That is the case in both Ethiopia and Malawi, which currently have among the lowest rates of electricity connection in the world, as highlighted by Figure 3.2.

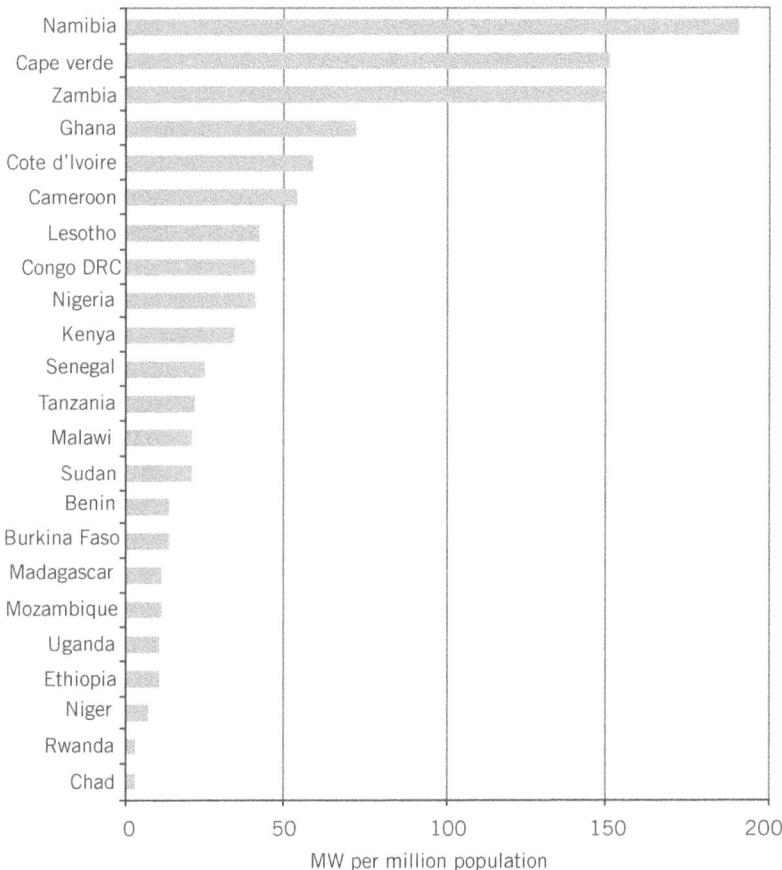

Figure 3.2 Generation capacity, megawatts (MW) per million inhabitants, 2005
Source: EIA, 2005

Subsidies directed to stimulating the consumption of fossil fuels present a specific set of drawbacks. Governments around the world spend many billions of dollars each year subsidizing both the consumption and production of fossil fuels, partially to alleviate energy poverty but also to increase domestic growth or to gain political support. However this leads to a number of unintended consequences, as follows:

- Economies become locked into long-term reliance on fossil fuels;
- Indirect support of unsustainable energy resources leads to various forms of environmental pollution, land degradation, and increased greenhouse gas emissions. According to the IEA, a complete phase-out of subsidies for fossil-fuel consumption would reduce CO_2 emissions by 5.8 per cent by 2020 as compared to business as usual (OECD/IEA, 2010);
- The distortion of market prices causes a chain of consequences that indirectly affects the economy, such as reducing investments available for cleaner energy production and other national priorities. Consumption subsidies amounted to $312 bn in 2009 (OECD/IEA, 2010) and production subsidies are estimated by the Global Subsidies Initiative (GSI) to be $100 bn annually (GSI, 2011).

Once introduced, subsidies are problematic to withdraw; not just due to the general discontent associated with rising prices, but because although low-income households typically receive a smaller share of the subsidies, they are disproportionately affected by their reform. For example, in Ethiopia, a reduction in kerosene and liquefied petroleum gas (LPG) subsidies resulted in the urban poor reverting back to using wood, charcoal, and agricultural residues for cooking, causing health problems, deforestation, and land degradation (Asfaw, 2012).

Many policy analysts recommend that policy reforms include compensation measures for the very poor who are most vulnerable to rising energy prices (Global Subsidies Initiative, 2011). The financing for this compensation can be effectively met through well-designed and targeted subsidies for on-grid energy systems; systems which focus on increasing connections and access to a range of energy services, rather than blanket price reductions which often do not help the very poor.

Despite their drawbacks, targeted financial mechanisms, including short-term subsidies, can be advantageous for on-grid energy delivery if aimed at assisting the development and market entry of innovative and beneficial technologies aimed at the poor. A more innovative subsidy model for on-grid energy delivery for the poor is a sliding scale, which provides the highest level of subsidy for the first initial amount of energy used. This can help the very poor access electricity for their most basic services, such as bulbs for lighting or mobile phone charging. Many countries, developed and developing, have adopted similar strategies, as highlighted by the South Africa experience in the 'Payments systems' section of this chapter.

Investments

Investment capital to build power plants can be sourced from private or public funding. According to an Energy Sector Management Assistance Programme (ESMAP) report, there are well over 700 electricity-generation plants in developing countries that have been financed and constructed (and are now operated) by independent power producers (ESMAP, 2008, ARE, 2011). Once an on-grid network has been set up, the funding of its extension can come from different sources, either through government support (alone or with support from international development banks and/or regional and national development donors), or private-sector investment, or through partial payments from the energy end users (households, businesses and community organizations).

Investments can come from various sources, both institutional and individual, with the expectation of receiving returns on investment. Common investment financings are explained in the following paragraphs, with loan guarantees, credit lines, and revolving funds being commonly used to finance on-grid connections to the poor.

Venture capital. This is high-risk financing with companies specializing in financing new companies or projects at the early take-off phase but with an expectation of high returns for the risks taken.

Debt financing. This is a common mechanism whereby a financial institution or investor provides capital and earns interest, but would not own shares of the project or company to which they lend. This type of financing is provided by both international and national commercial banks, multilateral development banks, debt investment funds, and private investors. The options include corporate or project loans under recourse or limited recourse structures, leasing arrangements, and full or partial guarantees. Financiers using this model specify minimum cash-flow generation projections, debt coverage, leverage, and other financial ratios for projects to qualify for such loans. Under such models, loans could be converted to some amount of equity ownership with a view to increasing the lender's rate of return. Debt financing is less risky as lenders must be repaid their investments first before any distribution is made to shareholders.

Equity financing. Financiers will be joint owners of the company, fund, or project, and financing is often long term with a preferential higher expected rate of return than debt as the financiers bear higher risks. The equity funds provide investment capital and expect, in return, a share of the project equity. The funders often have the rights to distribution after obligations such as tax are made. This type of financing could include joint venture partnerships, equity investment funds, pension funds, and venture capital.

Joint ventures involve shared ownership (of company and/or project) and are typically strategic alliances between two or more companies. This type of alliance could include provision of capital, technology, financial risk coverage, and skills sharing. Equity investment funds were created by international financial institutions and multinational organizations such as the World Bank and UN specifically to benefit the environmental and energy sectors. Examples are the International Finance Corporation (IFC) and the Global Environmental Facility (GEF), which provide such institutions with equity investment capital.

Revolving loan funds. These are structured to become self-sustaining after the initial capital investment. The fund is replenished by borrowers as they repay their debts, and the interest used to offset risks or manage the fund. From time to time, new loans are made to borrowers. The initial seed capital for revolving funds can be sourced from grants, government subsidies, or retained earnings. Many typical donor-led grants do not allow for funds to be used as revolving funds unless specified. These types of funds are often common for on-grid financing projects for the poor. See Box 3.9 as an example.

Credit lines. Financial institutions or government agencies dedicate lines of credit to particular market sectors, made available on a commitment basis, but returned if not used by customers. Depending on the nature of the financial institution, it can dedicate a line of credit to a specific client, for example to an energy entrepreneur wishing to purchase energy equipment and products. An interesting approach relates to bill financing, whereby loans are made through the national utility's customary household bill. An example is from Tunisia where a solar water heating (SWH) loan facility was set up as a joint initiative of the UN Environment Programme (UNEP), the Tunisian National Agency for Energy Conservation, and the Société Tunisienne de l'Electricité et de Gaz (STEG). Under this facility, interest rates and capital subsidies to customers were provided while STEG was utilized as a channel for recovering monthly loan payments. It has resulted in 7,724 systems being installed, with $5.7 m financed by three local banks and five qualified solar vendors (UNEP, 2011).

Carbon finance. In recent years, carbon finance markets have added a new dimension to the financing spectrum. Over the last decade, $27 bn of carbon finance has flowed to emission reduction projects, leveraging an additional $100 bn in related financing globally (World Bank, 2010), some of which has gone to on-grid electrification from renewable energy resources. As renewables have for a while generally not been able to compete financially with fossil fuels, particularly when the latter are subsidized, accessing carbon credits has been able to make some proposition more economically attractive.

Financing for entrepreneurs and consumer financing. In the context of on-grid energy systems, this type of financing mainly comes from the formal finance sector, the public sector, and/or through donor-funded programmes with a focus on financing. Banks can also provide a wide product range for funding appliances, such as asset finance loans to acquire machines, vehicles, and so forth; emergency loans if there is an immediate need that requires cash; and top-up loans added to an already existing one.

Formal banking has a higher level of security as funds are mostly safe from loss, theft, fire, misappropriation, and other risks associated with holding cash in hand. In addition, bigger loans are available and credit officers can often provide guidance and advice on technical areas such as business planning. However, formal institutions have stringent requirements for borrowing which many energy project developers, as well as consumers, may not meet. It is also important to note that if public utilities are bankrupt, financial institutions or investors are not willing to lend for a power purchase agreement from private developers (Hamilton, 2010). Donors, through banks, utilities, and NGOs, may also provide revolving funds or loan guarantees to offset the risks involved in lending for energy project development costs, and may lower lending rates. One such example is the Stima Loan in Kenya (see Box 3.7).

Box 3.7 Stima Loan credit facility in Kenya: financing lower-income customers

As part of Kenya's vision 2030, the government aims to increase electricity access to 40% by 2020. To reach this vision, Kenya Power and Lighting Company (KPLC) set a target to achieve 3 billion customers. The Stima Loan is aimed at the lower-income clients, including those living in rural areas. Through a loan from the French development agency, Agence Française de Développement (AfD), KPLC is targeting low-income families that cannot afford the high connection fees. AfD advanced KPLC 450 m Kenyan shillings (KSh) (approx. US$5.275 m) in June 2010 under the Stima Loan project. As of August 2012, the company had managed to connect 28,000 Kenyans to the national grid as a result of the funding. The poor people targeted by the scheme are expected to access electricity, and disbursement is carried out through KPLC or through Equity Bank and the National Bank of Kenya. Applicants are advanced between KSh35,000 and KSh100,000 repayable in instalments over 24 months. In mid-2012, at least 1,500 applications were being received every month, and the customer base had increased to 2.1 million.

In rural areas, customers pay 50% up-front while the balance is paid over a period up to 24 months. Under this scheme, beneficiaries pay no interest on the money advanced to them. In the peri-urban areas, customers pay 20% of the cost of connection up-front and are connected at preferential rates and conditions. The balance is paid in instalments over a period of up to 36 months. By early December 2012, about 37,000 Kenyans had benefited from the loan scheme with about 5,500 being from the western Kenya region. The region is sub-divided into 14 zones and disbursement is arranged based on the demand and population. While an identity card is required to secure the loan, connection is terminated on default of payment. The fund is a revolving fund facility, and as KPLC is not a credit institution, the higher the default rate, the lower the disbursement.

Note: Ksh1 = US1.15c as at 22 July 2013

Source: Njuguna, 2012; Mureithi, 2012; Nyabundi, 2012

A number of organizations have developed programmes to underwrite and reduce the risk of investments. The World Bank's Multilateral Investment Guarantee Agency provides investment guarantees to foreign investors in its member countries against non-commercial risks such as currency transfer and political instability. Export credit agencies of leading industrialized countries are also a major source of financing for large-scale on-grid energy projects.

Investments are closely tied to national policy developments and the availability of well-designed public risk reduction tools for commercial investment (Hamilton, 2010). Financing from the private sector is not instantaneous and a well-designed public finance policy will help it become established more quickly. For on-grid connections to assist the poor, public- and private-sector financing may be the most viable model as risks are absorbed by public finance through subsidies, or support for developers, which can then catalyse private financing.

Grants

Grants are often not closely associated with on-grid projects. This type of financing, with no expectation of a return, is often provided to projects or institutions that promote development. Many grants for on-grid projects are indirect and provided in the form of technical assistance by bilateral or multi-lateral agencies. They may include support for feasibility assessments, training (technical or business), business development, research and development, and marketing and awareness-raising, among others. A few grants are also provided to test out innovative models of delivering electricity to the poor in urban or peri-urban areas, some of which are described in this chapter.

Grants are sometimes combined with contingent loans to mitigate the risks of investments and financing. In this scenario, the loan element would be written off if certain conditions set by the donor were met. The GEF is one example of an institution that provides this type of loan, for example like the one provided to Cagayan Electric Power and Light Company (CEPALCO) in the Philippines to connect renewables to the national grid. This private utility built a 1 MW (6,500 solar panels on 2 hectares of land) PV power plant and integrated it into the 80 MW distribution network in the Philippine island of Mindanao. This loan will be turned into a grant once the plant has been successfully operational for five years. The GEF can also sponsor the up-front costs for project development, which can constitute up to five per cent or more of the total investment cost (GEF, 2009). However, funds through financing facilities such as GEF, the Public Private Infrastructure Advisory Facility, and through carbon financing require 'an inordinate amount of time spent preparing proposals and satisfying multiple financing windows' (Barnes et al., 2010).

One of the most common uses of grants is for capacity building, potentially one of the most important supporting services that is required to strengthen both the actors along the market chain and the enabling environment.

Box 3.8 lists key pointers for successful financing for on-grid delivery to poor or rural consumers.

Box 3.8 Key pointers for successful financing for on-grid delivery to poor or rural consumers

- Understand the clientele; financing terms must be related to willingness to pay.
- Financing cannot be provided on its own, and must be generated through awareness campaigns.
- Lending modalities must be simplified and made clear to consumers.
- Any appliance or product must be of a high standard ensuring consumer confidence. Successful energy access models show that formalized partnerships between utility companies and product suppliers are often needed to control quality and yield positive results.
- Repayment terms must be suitable for the customers – potentially more than two years. Customers in peri-urban or rural regions are dependent on seasonal incomes.
- The reach of financing institutions throughout the country must be significantly high if any on-grid connection for the low-income or rural areas is announced. Financing institutions' capacities must be built simultaneously with those of energy companies.
- Management and Information Systems would also need to be advanced so as clients can easily pay for the electricity.

Capacity building

With advancing technology and policy developments for on-grid energy delivery systems, many public- and private-sector personnel in developing countries will need technical support to be able to develop the most conducive enabling environment as well as ensure the required financial services are in place. A review of World Bank investments from 2000 to 2008 showed that about one-quarter of the bank's investment in energy access – about $1 bn – involved policy development and institutional building with funding for rural electrification master plans, policy frameworks, energy strategies, and heating sector reform (Barnes et al., 2010). Much of this investment went to support the development of public-sector capability to administer such projects.

The perception that an extensive market for on-grid electricity connections exists, is often accepted as a given, but it is common that many poor households are unable to connect because of high up-front costs or tariffs. There is therefore a lack of clarity concerning the demand management of such on-grid systems, to more accurately identify the true potential end users, as well as a lack of management capacity to more effectively target the unconnected end users, and to develop innovative strategies to try to help them connect. While financing is expected to be a major barrier for on-grid connection to the poorer households, financing institutions (banks, microfinance institutions, etc.) may need to be exposed to this type of lending, and users made more aware of the processes. Building capacity is therefore required for different institutional players on all aspects – technical, financial, and creation of an enabling framework. For example, government utilities and financial institutions working together on a grid connection model for low-income groups require each player to understand both the technical and financial settings, implications, and probable impacts.

A major gap is also the lack of sufficient skilled labour and industrial capacity to produce the required hardware and software needed to develop on-grid energy delivery systems which can deliver to informal settlements and areas located further from the energy production point. Often the lack of technical expertise leads to major inefficiencies in the delivery systems, as well as a reliance on imports of energy components and exposure to currency fluctuations. A strong, independent technical capacity is therefore essential for carrying out planning studies of grid extension to poorer households and rural areas; establishing standards and norms for electricity pylons, sub-stations, and household connections; building manufacturing bases; and designing regulatory frameworks for distribution grids, power plants and appliance production facilities. As many of the case studies have shown, local maintenance for the electricity services and appliances is also important, and therefore training to increase skills is very important in creating a successful delivery model.

Enabling environment

The development of a positive enabling environment for on-grid energy supply typically occurs through a series of factors that span a country's entire structure and include international dynamics and arrangements. For example, global and regional geopolitical trends – such as fossil-fuel prices and availability, or climate change international agreements – influence the economic performance of projects and consequently their viability. At the country level, the state of road infrastructure, the presence of trading channels (such as harbours and airports), the strength of national institutions, and the availability of natural resources (for example lakes, areas of geothermal activity, levels of solar radiation and wind) dictate whether some installations are possible or not. These factors are either impossible to change or difficult to influence in the short term; even if they lie outside the reach of single projects, they are important elements to take into account in risk analyses and implementation.

At the government level, there are several elements that can favour or hinder the provision of energy. Firstly, governments directly characterize the type of structure for energy provision in the country by adopting more or less public or private delivery systems. Government policies and strategies shape the way in which trade regulations and tax and tariff regimes are designed. This affects the import of fuel and technology and the development of local markets for energy supply and demand, including for appliances. For on-grid systems based on renewables, rights and allocation mechanisms for natural resources are particularly important. This latter set of factors is analysed in more detail in the following sections.

There are numerous examples which show that on-grid energy delivery systems are more effective at reaching the very poor when the state is actively involved in overseeing their delivery, for example, the Rural Electrification Collective Scheme (RECS) implemented by the Botswana Ministry of Minerals, Energy and Water Affairs, and the Botswana Power Corporation. This scheme was developed to promote and speed-up rural electrification rates in the country. Within the scheme, customers were required to form groups to apply for a connection, and to pay back the initial investment over a set period of months, depending on their preference (i.e. 18, 60, or 180 months). Between 1996 and 2003, access to electricity by rural households increased five-fold, with 80 per cent of new connections possible as a result of the scheme (SADC, 2010). Approximately 70 per cent of the electricity supply in Botswana is imported from Southern African Power Pool (the cooperation between the national electricity companies of each Southern African country, brought together under the Southern African Development Community – SADC), although plans under the scheme hope to implement renewable energy initiatives to reduce this number to 30 per cent by 2016 (DEA, 2011). A similar case in South Africa is outlined in Chapter 3, in the section 'Payments systems'.

The role of the state has also been evident in China where very high rates of on-grid electrification have been achieved, particularly with respect to grid extension to rural areas. The government has recognized energy access as a prerequisite for development and taken responsibility for its delivery, as highlighted in Box 3.9. It is important to note that this was achieved through significant financial support, including subsidies (and the combination of on-grid electrification as well as mini-grid energy delivery in very remote areas), as well as ensuring that supportive policies and their subsequent programmes were put in place.

National government policies

State actors play a crucial role in defining the enabling environment for on-grid energy delivery by shaping policies and regulatory frameworks, for example, helping to remove barriers to investment in large energy production or distribution schemes through incentives and standards. Over recent decades, the public sector has played a key role in delivering energy access directly, especially to rural or remote regions. This is important as in many developing countries, on-grid delivery systems are within the domain of the public sector.

Government policies and regulatory frameworks influence the choice of fuels, production, distribution technologies, and management systems. They can also help provide incentives to ensure that on-grid expansion reaches the poor, and can be accessed by them. This can be done through the bundling together of markets (wealthier markets with poorer ones, isolated markets with more centralized ones) into concessions which the private sector then bid for, ensuring they do not just focus on the more profitable and accessible areas.

Box 3.9 Rural electrification programme, China

Management: government bodies
Distribution: centralized
Scale: national (large scale)
Financing: government subsidies
Implementing organization: government

Despite China's rapid urbanization due to the last two decades of industrialization, 60% of its population still live in rural areas. The access to electricity is estimated to be 98.5%, which is extremely high compared to other newly industrialized or developing countries, particularly bearing in mind its national average per capita income of about $1,000. The high rural electrification rate is due to the favourable policies and programmes that the Chinese government has historically made in favour of electricity access for rural populations and their strategic development. The Government of China considers energy access to be a public service, and therefore it has provided continuous financial and political support. The users of the programme have contributed labour at all stages of the process.
    The Chinese rural electrification programme can be summarized in three stages:

1950 until the end of the 1970s
The driving force for rural electrification (particularly for small hydro) was the development of agriculture. Rural electrification was slow, yet impressive, and the progress was made under strict central planning with the investments and implementation led by the local governments and the counties.

From the late 1970s to the late 1990s
The Chinese government promoted rural industrialization, and emphasized the development of rural electrification by designing and implementing appropriate strategies and policies to suit local conditions and interests. This included strategies such as 'the one who invests, owns and operates' annual subsidies, to the implementation of small hydropower plants, plus special loans and other financing to meet objectives.

From 2000s to current
The turn of the century arrived with large-scale consolidation and upgrading of rural grids. The driving forces during this period were institutional reform, boosting electricity demand, integration of rural and non-rural electricity markets, fuel substitution to contribute to environmental protection, and an improvement in the quality of rural life.

    The significant achievements in rural electrification in China have been due to the clear role of the state in the provision of the electrical energy service and a strong institutional arrangement with clear roles for the different actors. The present Chinese government's plans are to have China's rural population fully electrified by 2015. For that purpose it has been implementing large-scale rural electrification programmes since the early 2000s, such is the case of the China Township Electrification Program launched in 2001 and completed in 2005. This was a scheme to provide renewable electricity to 1.3 million people in 1,000 townships in the Chinese provinces of Gansu, Hunan, Inner Mongolia, Shaanxi, Sichuan, Yunnan, Xinjiang, and Qinghai, and Tibet. Presently, this programme is being succeeded by a similar but larger one, the China Village Electrification Program, bringing renewable electricity to 3.5 m households in 10,000 villages by 2010. China is one of the examples of countries with explicit mandates for renewable energy for rural electrification.

Source: Jiahua et al., 2006

In urban areas, the problem of reaching poor people is often dependent on changing the mindset of urban utilities and often calls for special policies, investments, and innovative technical and financial solutions; infrastructure and capital investments for on-grid systems are often not the main issue in comparison with off-grid electrification systems (Barnes et al., 2010).

One example of a government-driven policy that has been successful in providing incentives for the development of decentralized on-grid energy systems is the small power producer regulations and feed-in tariff payment schemes in Thailand, as highlighted in Box 3.10. The policy of developing feed-in tariffs, especially for the development of renewable resources, is one where the government legally guarantees access to the power grid with a guaranteed price for power producers. Often, national utilities are then obliged to purchase any electricity generated at a fixed price. However, such policies do not necessarily bring down costs to the consumer as the price is set to be an incentive for private producers and can often be higher than the regular selling price of on-grid electricity – the incentive is set to enable a longer-term interest in green investments and to eliminate the negative environmental effects of fossil fuels. If the incentives are worthwhile, these independent power producers will also simultaneously provide power to the people who live in the vicinity where the power is produced.

Box 3.10 An enabling environment for decentralized grid power in Thailand

A successful example of government-driven policy providing an enabling environment that encourages private investment is in Thailand, where a series of policies has resulted in a dramatically increased contribution from decentralized renewable energy installations to the country's electricity supply. As of 2010, private enterprise had installed nearly 1.4 GW of grid-connected power, primarily from biomass installations, but also with a significant contribution from solar technology.

In 1992, Thailand introduced the small power producer (SPP) regulations, including a standardized interconnection system for the grid, and power purchase agreements (PPAs) for generators up to 90 MW (Greacen and Greacen, 2004). This was developed further in 2002 with the very small power producer (VSPP) regulations, which include a standard grid connection for small generators up to 1 MW (Greacen, 2007). These regulations fostered the rapid expansion of bio-refineries, primarily for sugarcane and rice; these small-scale generators enabled a variety of commercial opportunities, producing electricity, heat, ethanol, and food.

Prompted by concerns regarding reliance on fossil fuels, in 2006 the government introduced a feed-in tariff payment scheme, guaranteed for 7–10 years. The incremental costs have been passed on to consumers, although those using the least electricity (who are generally the least able to pay) receive subsidized rates. Other policy incentives for renewable energies included reduction or exemption of import duties, technical assistance, low-interest loans, and government equity financing (Yoohoon, 2010). The staggered introduction of policies and the use of feed-in tariff payment schemes have gradually led to widespread installation and diversification of renewable technologies in Thailand, and have greatly increased grid connection through a decentralized and affordable method for local populations.

Sources: IPCC, 2011; Greacen and Greacen, 2004; Greacen, 2007; Yoohoon, 2010

One of the most significant barriers to business growth in developing countries is the lack of a supportive investment climate in terms of strong governance and regulatory reforms (OECD/IEA, 2011). There is also often a weak implementation of the laws and regulations even if they exist. International banks or investors often only show interest when the policy regimes are strong and when they can get deals at the right scale (Hamilton, 2010). For example, to promote renewable technologies for on-grid systems, the provision of conditions for attractive risk-adjusted returns is a vital part of engaging interest from domestic and international private finance. Favourable foreign direct investment (FDI) policies are needed to ensure a long-term investment commitment, crucial to the renewable energy sector. In addition, to attract private capital into the on-grid energy sector, regulatory and incentive frameworks need to include separate regulations for the generation, transmission, and distribution of the energy, including the transparent announcement of future capacity targets and coverage for political risk (AGECC, 2010). A number of innovative regulatory investment programmes have been developed and are currently being promoted by major donors together with governments, to try to ensure a conducive investment climate, as highlighted in Box 3.11.

Box 3.11 Innovative regulatory investment programmes

Advanced market commitments

A key concept in catalysing private-sector investment for on-grid energy delivery systems is that of advanced market commitments (AMCs), which are 'temporary interventions to make revenues from markets more lucrative and more certain in order to accelerate investment' (Vivid Economics, 2009: i). AMCs have been piloted successfully in the health sector, encouraging the manufacture and distribution of vaccines for developing country markets. In the energy sector, feed-in tariffs and renewable obligations by governments are examples of AMC-based policy. These approaches could help the development of innovative 'demand-pull' measures that complement 'supply-push' measures e.g. capital grants (Vivid Economics, 2009). However, investments will only be stimulated if the energy suppliers respond, and thus the enabling framework must include a favourable climate for enterprise expansion and delivery. This can be achieved in a number of ways, such as governments introducing subsidies to achieve a stated outcome, to try to force consumers to behave in a specific way such as paying a set price, or directly purchasing the energy supplies for their direct consumption (Vivid Economics, 2009).

Results-based finance

Another concept, currently being promoted by major initiatives such as the Norwegian-led Energy+ Initiative, and multilateral organizations, is results-based financing (RBF). In essence, it follows the output-based aid approach whereby the 'transfer of money is conditional on taking a measurable action or achieving a predetermined performance target' (Knight, 2011). RBF mechanisms allow governments to tie resources to the results and reallocate if necessary, but it is essential that the private sector is active and has the right capacity to ensure they can deliver effectively to the end users. This is often a major issue in many countries, especially those where on-grid energy supply has traditionally been dominated by the public sector, particularly countries where governments are relatively unstable or consist of autocratic regimes. Thus, it is not only the policy formulation, but the larger political structure that determines whether energy enterprises will flourish or not.

Regulatory bodies

A number of regulatory bodies oversee the infrastructure construction of on-grid energy delivery systems, including environmental and agricultural departments which are particularly important for hydroelectric and bioenergy resources. Often the development of on-grid energy delivery systems falls under the control of more than one regulatory body (e.g. the requirement for collaboration between energy, environment, and forestry departments for the use of biomass technology), which requires significant levels of political cooperation to ensure that the systems are installed effectively. A lack of such coordination results in problems, either with the installation of the system itself, or its long-term sustainability.

In addition, national, regional, and international regulatory bodies all influence the method of on-grid energy delivery and its implementation, and must be aligned with each other. Regulations, management, and monitoring of implementing agencies and programmes, including critically assessing their success, are also important at the local level. An example of this from South Africa is outlined in Box 3.12.

Box 3.12 Regulatory bodies, South Africa

In 2009, 3.4 million households still remained unconnected to the grid (Mketsi, 2009) of which about half (1.7 million) were still classified as living in informal settlements. The South African government has prioritized the resolutions to the problems faced by the 1.7 million households living in informal settlements, including their connection to the grid, and this has been passed to the municipal councils to manage. Either the areas need to be formalized, or the households must move to other formalized human settlements before they can benefit from the national electrification programme. It is not cost-effective to electrify an unstable dwelling that may be demolished or destroyed by any climatic occurrence. For the 1.7 million households in formal settlements, a plan, including budget, is in place to ensure connection before 2014. A major challenge in the household electrification programme is the actual housing backlog, for which a special policy has been drawn up.

Source: Department of Energy, 2009

Regional government policies

On-grid energy delivery models are often not just governed by national energy policies, but also by regional energy policies, as is the case in several parts of Africa. On-grid electricity delivery models are influenced by groups of countries including the Common Market for Eastern and Southern Africa (COMESA), the East Africa Community (EAC), the Economic Community of West African States (ECOWAS), as well as the pan-African institutions of the African Union (AU), and the New Economic Partnership for Africa's Development (NEPAD). However, these intergovernmental organizations often focus on the delivery of very large electricity generation installations and regional power pools, and less on ensuring that the energy is available and can be accessed by the poor.

For example, the Inga Dam in Congo, is a priority project for several African development programmes and is expected to produce more than double the power of the Three Gorges Dam, enabling the DRC to increase its energy exports to surrounding countries. An agreement has already been signed with South Africa to more than triple the amount of electricity the DRC currently supplies, a large proportion of which is expected to power South Africa's mining industries rather than poor households (International Rivers, n.d.), as highlighted in Box 3.13.

Box 3.13 Investment in hydropower in the Democratic Republic of Congo (DRC)

The Congo River in the DRC is the location of the world's largest proposed hydropower facility, the Grand Inga Dam, as well as two currently operating dams (Inga 1 and 2) and a third under construction (Inga 3). The proposed fourth dam would be the largest construction along the river, and could potentially generate 39 GW of power from 52 turbines. With only 6% of the DRC's population having access to electricity, and 900 million people throughout Africa without access, there is a desperate need to increase generating capacity on the continent.

The Grand Inga Dam is expected to cost at least $80 bn, and requires significant external investment. The project is being driven by a variety of intergovernmental organizations and consortiums, including the New Partnership for African Development (NEPAD) and the Southern African Development Community (SADC). Funding for the project is expected to come from several multilateral investment banks (with some having confirmed their support); the World Bank, European Investment Bank, and the African Development Bank (AfDB) are some prominent examples. Public and private investment is being sought for a mixed PPP model.

In addition to the Grand Inga and Inga 3, there is currently a project being undertaken to rehabilitate the Inga 1 and 2 Dams, using funding from the World Bank, European Investment Bank, and the AfDB. The dams were constructed in 1972 and 1982 respectively, and have a legacy of debt that has been one of the driving factors in DRC's spiralling economic crisis.

One of the primary concerns regarding the project is that much of the power generated will not benefit local populations; many people displaced by the original dams have never been recompensed, and as of yet there have been no plans announced to increase grid access, although the Congolese government has signed an agreement with the South African government to implement and run the infrastructure through national energy companies. The large amount of foreign investment involved is expected to result in the additional generated power primarily benefiting industry and more well-off sectors of society.

Sources: International Rivers, n.d.

Although many national, regional, and cross-national programmes fail to address the needs of the poor, some strategies are formulated specifically with development of the poorest in mind; an example of this is the new regional 'Energy Access Strategy' being implemented by the East African Community (EAC) in partnership with UNDP and the German Agency for International Cooperation (GIZ), which recognizes the need to develop pro-poor regulatory frameworks of on-grid energy delivery. One of the policies being pursued is the 'East African Power Master Plan', which seeks to develop common energy policies across the East African region to encourage private-sector investment in grid-based energy services. The strategy hopes to map the energy service needs of half the region's population by 2015, and foster close relationships between energy stakeholders (e.g. ministries, enterprises, NGOs, donors, and utilities). A key figure is that poor households in the region are estimated to currently spend approximately $1.2 bn on meeting their energy requirements, which could provide a substantial finance incentive for on-grid utilities if they can capture these end-user payments (EUEI-PDF, 2008).

Inter- and intra-governmental collaboration is also important to make electricity generation and its trade between countries more successful. In Africa, regional power pools such as the Southern African Power Pool have been created to provide reliable and economical on-grid electricity supply to the consumers of the member countries. Similarly, if the Millennium Development Goals (MDGs) are to be met through on-grid energy delivery, government ministries within countries, and between countries, will need to better coordinate. The energy sector will develop faster if the infrastructure for on-grid energy delivery is in place, bringing economic improvement into regions through productivity increases resulting from increased access to energy. In addition, increased access to energy can also boost health through the provision of water for drinking and improved sanitation facilities in underserved regions, two of the most important indicators for tackling poverty.

Social context

Energy from on-grid delivery systems predominantly benefits the areas of society that are physically easy to reach (urban, peri-urban, and surrounding rural areas). The on-grid systems are structured to serve the end users who can afford to pay (particularly those who can afford to connect to the grid when it passes close to them), and operate within established financial markets. Poor households continue to remain locked in energy poverty, due to the financial barriers to grid connection. This restricts their ability to improve their livelihoods, unless specialized subsidy or tariff structures are developed by or with utilities, both for the initial connection and the ongoing payments.

Key aspects to consider in developing more appropriate on-grid energy systems include assessing needs, opportunities, and market demand; maximising local benefits in the market chain; and designing appropriate payment systems to ensure financial sustainability, such as the challenge of overcoming the payment of connection fees of on-grid systems. Although on-grid energy delivery systems are generally managed within central urban areas, their energy processing plants are often situated away from urban areas, for a variety of social, logistical, political and geographical reasons, including the availability of cheap lands, and the location of suitable sites for hydro dams. As the majority of the customers are therefore located some distance away from the on-grid system designers and project managers, the local adoption and acceptance of the systems is sometimes not one of their highest priorities – which needs to be addressed.

Historically, on-grid energy systems have been implemented to supply power for industrial production or export and have commonly failed to increase access to local communities, sometimes even leading to increased poverty and social conflict due to the negative environmental effects of the project on the local area (Behrens et al., 2012). Large-scale hydropower is a common example (Namy, 2007); dams can produce a large proportion of a country's electrical power supply, but often with significant negative effects on local populations that will not necessarily benefit from the project's improvement in energy access. Cross-border issues are increasingly becoming an issue for large dam projects such as within the Nile Basin, where a number of countries are ultimately affected by one country's actions. Additionally, the increasing use of arable land for cultivation of energy crops for exported biofuels is leading to resource conflict and land degradation in many countries, particularly in East Africa (Deininger et al., 2011).

The success of any state intervention on on-grid energy delivery models depends on the level to which priorities are harmonized, and the level of community 'ownership' or 'buy-in'. Unlike mini-grid and off-grid energy delivery systems, which are often specifically designed with the user in mind, if on-grid energy delivery systems are going to be successful at reaching the needs of the poor, the active buy-in from these poor end users is critically important, and a move away from a purely top-down approach is required. The customers' willingness and ability to pay for energy services is one such example, and needs to be assessed as a business priority of the energy utilities. In some circumstances (where electricity has been heavily subsidized in the past or where it is seen as part of the state's assistance duties) people might have false expectations regarding charges, and default on payments. In other cases, customers might not be in a condition to sustain the cost of the service or that of appliances that make access to an energy connection useful to them. Thus, consultation with, and participation of, consumers from the very first stages is important, rather than designing systems first and then trying to engage the end users (which is more often the case). This is particularly the case with respect to the designing of tariff payment systems, and providing a satisfactory maintenance service.

CHAPTER 4
Mini-grid energy delivery

Mini-grid systems are often the most suitable energy supply solution for very remote, rural areas that cannot be reached by grid extension. Mini-grids make use of local environmental resources as well as fossil fuels, and depend on the technical, managerial and financial capacity of the communities they serve. A common challenge for many systems set up with the help of donors, but managed by the local community, is how to achieve long-term sustainability. This involves maintaining equipment and retaining technical expertise locally, in areas where often the long distances to urban centres to seek replacement parts and technicians is a real challenge. This chapter looks at examples of mini-grid systems where these challenges have been addressed.

Keywords: decentralization, low carbon, mini-grid, public–private partnerships, remote areas, small enterprise management

Interest and support for mini-grid energy delivery systems is gradually increasing. The International Energy Agency (IEA) states that while grid extension is the most suitable energy delivery method for all urban areas (and approximately 30 per cent of rural areas) it is not cost-effective for the remaining 70 per cent remote, mountainous, and sparsely populated areas which are often subject to high technical losses through long distance transmission lines (OECD/IEA, 2011). It is expected that to meet the UN's goal of energy access for all by 2030, approximately 65 per cent of the areas not covered by grid extension will receive their energy from mini-grid systems (with the remaining 35 per cent provided by stand-alone off-grid solutions). To achieve this scenario, the IEA predicts that US$12.2 bn of annual funding must be invested in mini-grid technologies up to 2030 (OECD/IEA, 2011).

Mini-grids are often the most suitable energy supply solution for remote, rural areas, in terms of affordability, effectiveness, and efficiency. Using local resources (e.g. solar, wind, biomass or biogas), but also fossil fuels (e.g. diesel), decentralized mini-grid systems provide a more economically feasible and sustainable alternative to grid extension (Yadoo, 2012; Yadoo and Cruikshank, 2010).

There is growing acknowledgment from the international community that mini-grid connections will have a vital impact on education, health, and livelihoods, and that together with off-grid solutions, will play an essential role in reducing energy poverty in areas where on-grid systems are not able to reach. Donors are increasingly focusing on developing programmes to increase the technical and managerial capacity of local communities to install and

http://dx.doi.org/10.3362/9781780447612.004

manage mini-grid systems, as well as the government institutions to oversee them. Local private-sector initiatives are also increasingly being recognized as fundamental actors in all types of energy delivery systems (including mini-grids), alongside cooperatives and civil society organizations.

Energy market chain

Mini-grid energy delivery systems rely on localized, small- or medium-scale power sources, less than 3 MW in size (although typically much smaller), supplying a local distribution grid connected to nearby domestic, private, and institutional customers. Energy produced by mini-grids is distributed in the form of electricity, gas, or in some cases mechanical power. By definition, these systems are not connected to large-scale or national grid supply, with their management and maintenance being confined mostly to the local authority, community, or private company. They are generally more flexible systems than nationally managed ones, and utilize a range of technologies to provide energy services in areas where the national grid cannot reach or is unlikely to reach in the near future (for physical, technical or financial reasons). When electricity is generated, mini-grids typically do not incur the same high percentages of losses as national-scale grids, making distribution potentially more energy-efficient, although the cost of production (per kWh) might well be higher due to the reduced economies of scale. If properly planned, mini-grids can be constructed so they can easily be integrated should the grid reach the region and become a decentralized input to the national network. Figure 4.1 provides an example market map of a solar PV-powered mini-grid system developed by Sunlabob in Laos, which aims to sustainably deliver electricity to remote poor households in the country. Further details on this model are provided in Box 4.4.

Energy sources and technologies

A number of energy sources can be utilized within mini-grids systems, ranging from fossil fuels and diesel-powered generators, to small-scale renewables (micro-hydro, solar photovoltaic – PV, wind, biomass, biofuels, and possibly geothermal) as well as hybrid systems such as diesel and pure plant oil-powered multi-functional platforms as supported by UNDP in West Africa (UNDP, 2012). The technology implemented often consists of generators (for diesel, biodiesel, and pure plant oil) and micro stations (biogas digesters, biomass boilers and gasifiers, solar farms, hydro dams, run-off river systems, and wind turbines). Hybrid systems have garnered particular interest due to their combining of 'dispatchable' sources on-demand (e.g. diesel fuel) with 'non-dispatchable' sources (e.g. wind), which can result in increased reliability and load matching (GVEP International, 2011). The importance of smaller energy production centres is also linked to energy security as a portfolio of options, and a diversified mix of energy supply sources, creates a safety net of numerous, relatively small units with a low individual cost of failure. This increases the overall resilience of the system.

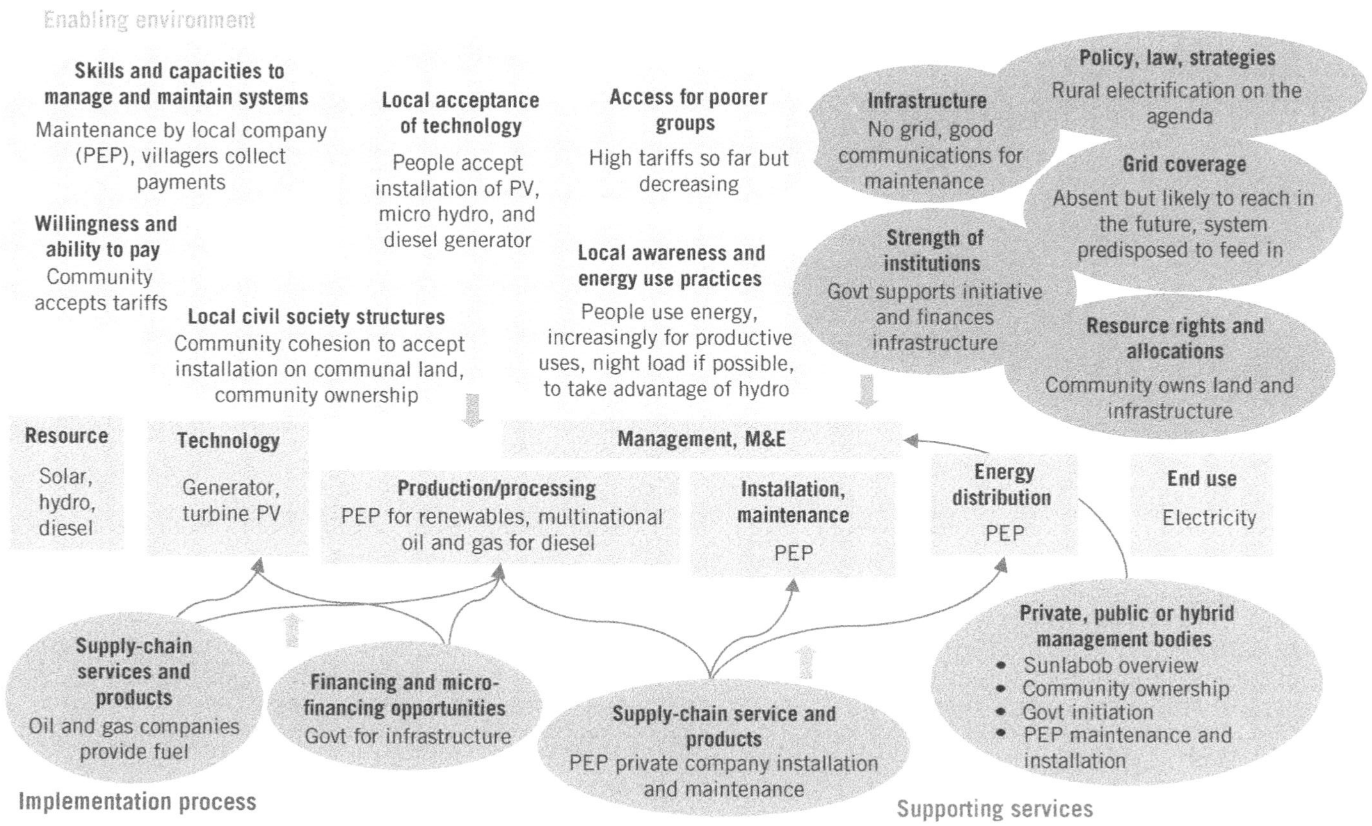

Enabling environment

Skills and capacities to manage and maintain systems
Maintenance by local company (PEP), villagers collect payments

Willingness and ability to pay
Community accepts tariffs

Local acceptance of technology
People accept installation of PV, micro hydro, and diesel generator

Local civil society structures
Community cohesion to accept installation on communal land, community ownership

Access for poorer groups
High tariffs so far but decreasing

Local awareness and energy use practices
People use energy, increasingly for productive uses, night load if possible, to take advantage of hydro

Policy, law, strategies
Rural electrification on the agenda

Grid coverage
Absent but likely to reach in the future, system predisposed to feed in

Resource rights and allocations
Community owns land and infrastructure

Infrastructure
No grid, good communications for maintenance

Strength of institutions
Govt supports initiative and finances infrastructure

End use
Electricity

Resource	Technology	Production/processing	Management, M&E		Energy distribution
Solar, hydro, diesel	Generator, turbine PV	PEP for renewables, multinational oil and gas for diesel	Installation, maintenance	PEP	PEP

Supply-chain services and products
Oil and gas companies provide fuel

Financing and micro-financing opportunities
Govt for infrastructure

Supply-chain service and products
PEP private company installation and maintenance

Private, public or hybrid management bodies
• Sunlabob overview
• Community ownership
• Govt initiation
• PEP maintenance and installation

Supporting services

Implementation process

Figure 4.1 Sunlabob solar mini-grid market map from Laos
Note: PV = photovoltaic; PEP = private energy provider

Unlike on-grid energy delivery, mini-grids have the potential to be tailored to exploit the environmental resources of a particular geographical area, and better meet the requirements of the customers by choosing technologies that directly solve a particular need (e.g. a hydro scheme that makes use of a local river to provide pumping power for irrigation). The delivery model can also be designed to suit the particular circumstances in terms of prices, payment schemes, and type of services. Mini-grids that are powered by renewable energy resources present the additional advantage of being independent from external price and supply availability factors, which often affect fossil fuel-based electrification systems. There is currently very little information on the total proportion of mini-grid energy systems that are supplied with renewable energy, but it is believed to be quite low, and a tiny fraction of its potential.

The increasing shift of international policy focus towards a low-carbon development pathway for developing countries is playing an increasing role in prioritizing the use of renewable energy, often within mini-grid energy delivery systems, to help countries bypass their reliance on carbon intensive energy production. It should be noted that although many developing countries have contributed very little to historical carbon emissions and currently still produce relatively very low levels of greenhouse gases, a lot of international funding is focused on low-carbon development which is pushing this agenda.

Ownership and management

Mini-grid systems can be maintained either by organizations representing the local community or by private external agents. Mixed situations are also common in which partnerships are formed between private companies and public organizations to share responsibilities in the construction, management, and maintenance of systems (e.g. Husk Power in India). The implementation of mini-grid systems has been facilitated by regional governments, NGOs (multilateral and unilateral donors), social enterprises, and partnerships with civil society organizations, as outlined in Box 4.1. The involvement of NGOs is often focused on the development of the market, informing government policies, raising awareness, and providing supporting services, including capacity development and training. The systems often also involve some level of public-sector direct support, although most government support comes indirectly via policies and subsidies.

A review of projects and studies has shown that there are many options for designing and managing mini-grid systems. Although a wide range of different models have been piloted in energy access programmes (with some more appropriate or cost-effective than others), no ideal model has so far been identified, due to a range of cultural, environmental, and socio-economic factors. The management of the systems can be carried out by private companies in partnership with community organizations, through NGOs and with public support. In Nepal, many micro-hydro systems are also owned, managed, and

---

Box 4.1 Management approaches

Community-based model: this has been tried out extensively around the world with varying levels of success depending on the level of community participation and pricing policy. In this model, the community is the owner and operator of the system and provides maintenance, tariff collection, and management services. These models tend to be more common in developing countries where the private sector or utilities remain limited.

Private-sector operator: in this case, a private actor establishes the mini-grid system, although there may be many different implementation and management structures based on whether ownership is retained or passed on to another actor. The participation of the private sector is only realistic if the project is economically attractive.

Utility-based operator: this is the most common approach for rural electrification in most developing countries. In this approach, a utility company is responsible for all or part of the electricity system. The company may be a state institution, private individual, or community cooperative.

Hybrid business model: these are diverse models and may involve different entities owning and operating different parts of the system. For example in this case, a utility or a private company might own and operate a system and a community cooperative manage it on a daily basis. In addition, a private company would provide the technical backup and management advice.

---

maintained by private entrepreneurs who initially had financing through a development bank, from government, or from donor subsidies.

In a number of countries, mini-grid systems are leased by a public utility to a private-sector actor, or are run by a cooperative. The management and maintenance of mini-grid systems is often transferred to private enterprises and/or community organizations once the installation has taken place. This is outlined in Box 4.2 which explains the small enterprise management model first piloted in Peru. This model has also been replicated in other countries including Zimbabwe, Mozambique, Malawi, Nepal, Nicaragua, and the Philippines by organizations such as Practical Action, Blue Energy, Sibat, and GIZ's Energising Development programme. In these schemes, either local entrepreneurs or community members (through a community initiative) are trained to take over the long-term management of the systems.

These models have the potential to provide significant benefits for the communities involved (e.g. increased income-generating capabilities from enhanced skill levels) through high degrees of participation and interaction.

A common challenge for many mini-grid systems that have been set up with the help of donors, but are managed by the local community, is how to achieve their long-term sustainability. This challenge includes the maintaining of the physical equipment of the systems, as well as retaining the technical expertise in the communities, as highlighted in Box 4.2. Often the long distances to urban centres, where replacement parts or technicians are more accessible, make maintenance a real challenge. The potential for this issue can be addressed by focusing on the development of appropriate management models and financial payment schemes (including maintenance funds), taking into account the internal social relations of the community, and its

---

Box 4.2 Small enterprise management model

The small enterprise management model was designed and implemented by Practical Action, following a post-implementation study of the performance of off-grid generating schemes installed in Peru during an Energy Sector Management Assistance Programme (ESMAP)/Practical Action project between 1996 and 1998. The model was initially piloted in the village of *Conchán*, in the northern Andes of Peru, and has since been implemented, adapted, and employed in other contexts in other areas by Practical Action, embracing all village-level issues that typically affect the sustainability of mini-grid energy systems.

The principal objective of the model is the efficient management of small-scale stand-alone mini-grid electricity systems. The model covers all physical elements related to the generation and distribution of electricity, including the energy resources, civil infrastructure, electro-mechanical equipment and the electricity transmission (when needed), distribution networks, and household connections. Management refers to all the activities involved in sustainably operating, maintaining, and administrating these small-scale electrical systems.

The model has introduced the concept of 'entrepreneurial management', which is generally a new idea in remote rural areas. The model also promotes financial management free from the interference of political interests, either from within the community or outside, and with the appropriate number of people involved in the operation, maintenance, and management. This has had positive consequences for the financial performance of the systems, as sustainable businesses providing energy services. The model promotes and supports the rational use of energy and its use for income generation through the provision of local services and the transformation of products, leading to the promotion of community development. The participation of the population is promoted as much as possible in the planning of the management system and in decision-making about the operation, maintenance, and management of each system. This ensures that the end users assume responsibility for the system and recognize the rights and obligations of the different actors involved.

There are four main actors involved in the operation of stand-alone electricity systems: the owner, the micro-enterprise operating the service, the users, and the users' auditing committee. In exchange for the enterprise's efforts, the owner pays a monthly sum, with the rules on payment collection from users and the use of this money being established as part of the tariff structure. The amount to be paid is defined and agreed upon and stated in the contracts, in order that both parties adhere to it. The contract also states the rules on disputes and considers how these can be resolved, taking into account relevant local laws where necessary.

Source: Sanchez, 2006; Yadoo, 2012

---

organizational structure, values, and capabilities at the group and individual level (as mentioned in Box 4.1).

However, despite these difficulties, there are some examples of mini-grid systems which have been able to overcome these long-term sustainability issues. One such example is the civil-society initiated mini-grid energy delivery system that was developed without the direct involvement of public- or private-sector actors in southern Brazil – the Creluz cooperative. This is described in Box 4.3 and highlights how an innovative financing system was developed to overcome the barriers of maintaining systems, as well as scaling up success.

Box 4.3 CRELUZ micro-hydro cooperative, Brazil

The CRELUZ cooperative was formed in 1966 by residents of the Rio Grande do Sul state in southern Brazil, and was initially set up as a small electricity distribution network with power supplied by the national grid. This initial investment in the first micro-hydro plant led to the establishment of a profit-making enterprise, with income used to cover the initial costs and to fund additional hydro-installations in the area, through mini-grid delivery systems.

As of 2009, the cooperative owned and maintained a network of six micro-hydro installations and 4,500 km of power lines, which supplied approximately 27% of the required electricity for 20,000 families in 36 municipalities; the remainder is supplied from the national grid. Each family paid a subsidized tariff per kWh of electrical power, with the usage supplied by a meter reading. The cost of the electricity is approximately the same as conventional on-grid supply, but the profit made by the cooperative is invested back into expansion of the mini-grid and social programmes.

A sliding-scale of tariffs means that poorer families receive a discount of up to 64% off the average price of $0.20 per kWh, while the worst-off receive free electricity. This has helped to bring much of the population into more involved roles in the local society, improving livelihoods and increasing income for many families who can then move higher up the tariff scale. The hydro-dam installations have been designed to minimize environmental impact, and profits made from the cooperative are funding other community initiatives such as rainwater harvesting systems. The system is currently predominantly self-contained in terms of market actors, with only minimal support coming from external services (both public and private). The deficit in electrical power supply is bought from the national grid, however the cooperative is considering external finance options to fund a further expansion of operating capacity.

Source: CRELUZ, n.d.

Figure 4.2 gives an outline of the energy service company (ESCO) model, which can be used to oversee the installation, management and maintenance of mini-grid energy delivery systems in coordination with the other main actors. An ESCO provides a vital link between the grid utility, the end users, and their management structure (e.g. a village committee). As a national-level utility is often focused on large-scale grid systems, they are reluctant to become directly involved in relatively small-scale mini-grids, which is where the ESCO comes in, providing the critical role.

Public–private partnerships

Public–private partnerships (PPPs) are expected to become an important model for energy delivery through mini-grids; typically with the capital start-up costs being covered by government investment (in part or in whole) through multilateral and bilateral support loans, while the ongoing maintenance costs are financed by local companies and/or private energy suppliers. The Alliance for Rural Electrification (ARE), is a leading institution working to fill the knowledge gap around how renewable energy technologies can provide sustainable and financially viable energy access for developing countries. ARE has released a number of publications covering various aspects of renewable

Government, World
Bank, ADB, NGOs

• Builds permanent electricity distribution infrastructure in the village
• Continuously educates villagers on sustainable use of electricity

Power purchase agreement (PPA)

Grid utilities ← Energy service companies (ESCOs)

• Sets up and connects energy generation equipment
• Sells electricity (kWh) to the village committee
• Trains village technicians to operate and maintain the local grid

Villagers

Sells electricity to the villagers and collects fees

Village committee

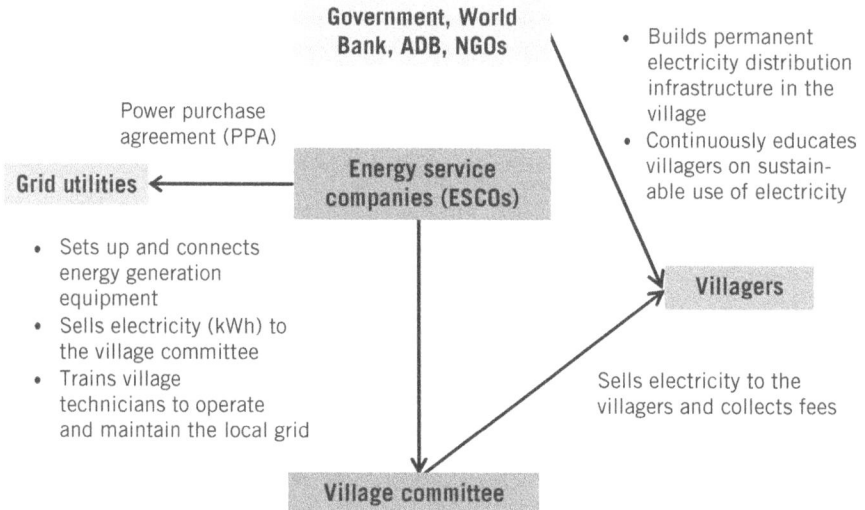

Figure 4.2 The ESCO model
Source: GVEP International, 2010

energy technologies for mini-grid energy delivery models, including business models, technologies, and policies, particularly with respect to PPPs. The publications outline the following key recommendations from completed and ongoing mini-grid projects:

• Regulation has to be an instrument favouring new projects and not be a burden. It needs to be light and flexible for the small power producer in terms of standards and tariffs. It also has to protect rural consumers.
• Power purchase agreements (PPAs) must be fair and binding to protect every actor equally. They should be as standardized as possible to decrease administrative costs, increase efficiency, and simplify procedures. They should also be signed over a longer period of time, be flexible, and be revisable when it comes to tariff.
• PPAs should be indexed to foreign exchange rates to offset the devaluation risk and encourage long-term investments from foreign sources.

Figure 4.1 and Box 4.4 illustrate an example of a PPP in Laos where the success of the initiative came from the collaboration between the actors, which resulted in the building and maintaining of a sustainable financial structure for delivering energy. This project is evidence that successful implementation of mini-grid systems can be a precursor to national grid expansion.

Payment systems

In a similar manner to large-scale on-grid systems, users typically pay for services at regular intervals (e.g. monthly billing), or through a pre-paid

---

Box 4.4 Hybrid mini-grid in Laos: a public–private partnership venture

Sunlabob is a private commercial company in Laos that has installed a hybrid mini-grid, operated by a PPP. Public partners funded the infrastructure and fixed grid (which is owned by the village community), while Sunlabob formed a local private energy provider (PEP), which owns and finances the moveable assets (power generation system). The PEP employs local villagers to collect the tariff fees, and is responsible for the maintenance of the mini-grid. Without ongoing public financial support, the electricity has remained relatively expensive for users, although rates of return are slowly increasing as local revenue-generating activities develop. Through this project, Sunlabob has committed itself to a 25-year PPA with the village, giving it an internal rate of return of around 15%.

The hybrid system combines a 12 kW small hydro generator with a 2 kilowatt peak (kWp) PV system and a 15 kilovolt-ampere (kVA) diesel generator. It supplies 105 households with a daily peak load of 8 kW; this peak remains for 3–4 hours per day. The system runs almost entirely on renewables, as the energy produced from hydro and PV is enough to cover the majority of the households' everyday needs. Additionally, since hydropower is available for night loads the costs have been significantly reduced. Solar PV was added for the dry season, with the diesel generator as back-up to address any unexpected demand.

The system has the potential for numerous benefits; the infrastructure of the hybrid grid means that connection to the national grid would be relatively simple and inexpensive. By adding the system to the national grid, lower tariffs could be offered for the local users, revenues could increase for the PEP, and the regional utility (grid operator) could benefit from increased generating capacity. It is estimated that by connecting the mini-grid to the national grid, the cost per kWh of electricity to local end users will be cut by approximately 75%. It also means a switch to a utility-based management model in the process. The support schemes demonstrated here have been conducive to long-term private-sector involvement, and have the potential to form a link between mini-grid and national grid energy delivery.

Source: ARE, 2011

---

metering system, covering the ongoing costs of the resources, management, and maintenance. However, many installations are still in a pilot phase where payments do not cover the installation costs – these have usually been covered by grants from donors to pilot the concept. To further up-scale these systems, innovative financial mechanisms need to be designed to ensure that the installation costs, as well as ongoing management and maintenance costs, can be paid for on an ongoing basis, rather than through further subsidization.

Among the new models that are being developed are Practical Action's Mulanje Electricity Generation Authority (MEGA) initiative in Malawi, GPower's micro-energy systems in Kenya, and the Small Scale Sustainable Development Fund's (S³IDF) initiatives in South Asia. These models try to address the financial sustainability challenges by setting up small micro-grid companies that can run micro-grid systems as profitable enterprises. In these cases, profits are re-invested to extend the systems or to replicate them in neighbouring communities. The CRELUZ cooperative described in Box 4.3 is an example of a civil society-based organization which operates a tariff scheme for its customers and subsidizes the poorer customers from the payments made by the more affluent customers. CRELUZ's sliding scale of payments is used to allow poorer members of the community to access the energy markets by providing them with a discount, known as a 'lifeline subsidy', or in the case of the poorest, providing free electricity. The theory is, that over time the expected improvement in the livelihoods of these poorer households will allow them to start increasing their tariff payments. Such schemes are based on users' ability to pay rather than expecting the same level of payment from all income groups. As with the CRELUZ cooperative, other mini-grid systems cover their maintenance costs through the payments made to the owners of the systems (either private-sector actors or community organizations).

Other mini-grid systems involve a level of subsidization by NGOs, multilateral or unilateral donors, and others, predominantly to cover the initial investment and capacity development costs, but also occasionally to reduce the ongoing payment schemes. One example of this is Senegal's 2007 programme entitled, 'Designing Technology-neutral Concessions for Rural Electrification'. This programme was funded by grants from the International Development Association and the Global Environmental Facility (GEF) which were used to cover subsidies under an output-based aid approach (de Gouvello and Kumar, 2007). Certain community-level mini-grid and off-grid energy delivery systems, such as biogas digesters for community services, have specific maintenance and payment schemes, as highlighted in Figure 4.3. The figure indicates the interaction between the various organizations that initiated the scheme; the municipality that provided overall support and seed funding; the local community enterprise which manages the maintenance and payment; and the community committee which oversees the community involvement in the scheme. The effectiveness of the interactions between all these stakeholders is critically important in ensuring whether such mini-grid energy delivery models are sustainable or not.

An example of a community-based management and payment system for a mini-grid energy delivery model based on wind energy resource is summarized in Box 4.5. Interestingly this is a hybrid system, which involves a number of off-grid wind turbines but utilizes a mini-grid management and user payment scheme. The energy delivery model technologies were designed to delivery energy to individual energy users, ensuring that the owners take full responsibility for solving any technical issues and also reducing the cost of supplying the energy to individual households (this is a cost which can be prohibitive for mini-grid schemes supplying highly dispersed households). However the management and payment systems were designed in the style of mini-grid schemes, ensuring that the skills for the maintenance of the systems were kept in the community, and one community company was responsible for the installed units.

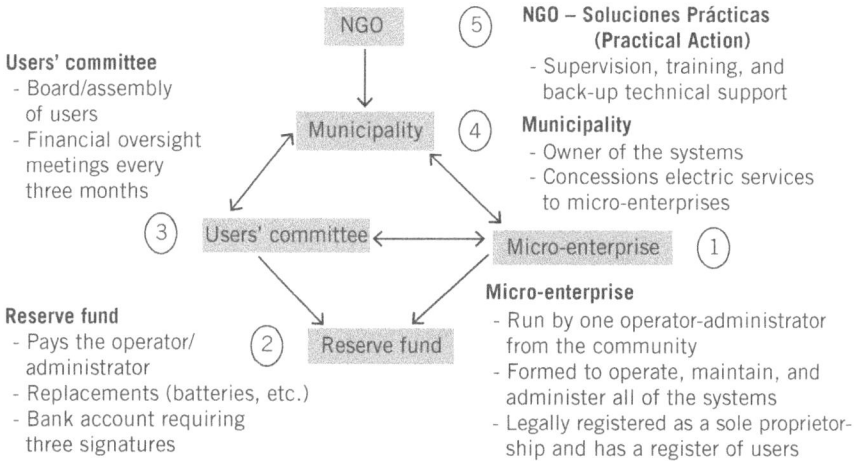

Users' committee
- Board/assembly of users
- Financial oversight meetings every three months

Reserve fund
- Pays the operator/administrator
- Replacements (batteries, etc.)
- Bank account requiring three signatures

NGO – Soluciones Prácticas (Practical Action)
- Supervision, training, and back-up technical support

Municipality
- Owner of the systems
- Concessions electric services to micro-enterprises

Micro-enterprise
- Run by one operator-administrator from the community
- Formed to operate, maintain, and administer all of the systems
- Legally registered as a sole proprietorship and has a register of users

Figure 4.3 Community-level mini-grid biogas system
Note: The figure highlights the interactions between the various organizations involved in a biogas mini-grid system, from the micro-enterprise set up to manage the scheme, the users committee, and the local municipality, through to the NGO which initiated the scheme

---

Box 4.5 Community wind generation project, Peru

This project, coordinated by *Soluciones Prácticas,* the Latin American arm of Practical Action, involved the installation of a series of wind turbines for 33 houses and two community buildings in *El Alumbre,* an isolated rural village in the northern Peruvian Andes, between January 2008 and January 2009. It was originally planned to be a mini-grid connected to a few larger turbines, which would supply electrical power to the buildings in the village. However this was deemed too expensive due to the large distances between some houses. Instead, each building received its own turbine; while this was technically an 'off-grid' installation, the management and payment system is indicative of a typical mini-grid set-up.

A collaboration between NGOs and the municipal government provided the funding to purchase the installed equipment, and in conjunction with the local population organized a micro-enterprise and user committee (composed of the entire village). The micro-enterprise consisted of an elected individual from the village, who was responsible for the maintenance and administration of the turbines, and collection of a monthly tariff paid by each household. The tariff was set at a price below what the villagers were generally paying for their traditional fuel supplies (kerosene and candles) each month, providing a saving for them. The money raised was used to pay the administrator of the micro-enterprise, and to keep a reserve fund (controlled by the committee) for purchase of spare and replacement parts. The technology itself was kept under the ownership of the municipality, which shared responsibility with the reserve fund for the upkeep of the installation.

The extensive inclusion of the villagers in the management model, and the creation of a micro-enterprise, encouraged the development of local skills, and developed the concept of energy generation and distribution as a service among the community. Their participation in the management model has helped the ensure the sustainability of the system.

Source: Ferrer-Martí, 2012

Supporting services

Mini-grids need a particular set of supporting services, especially in developing country contexts. These are financial services (in general the required financial services are not available from the existing banking system); technical support for the ongoing maintenance of the systems; and specific capacity building to ensure that the businesses that run the mini-grid systems are able to deliver a service that meets the needs of the end users.

Financial services

Financial structures and support for mini-grids are more diverse than on-grid energy delivery systems. Different financial models may also be applied to cover up-front capital costs and the ongoing operation, and maintenance costs. Building sustainable financial structures can be challenging despite a growing understanding of barriers and an increasing number of financial tools. While mini-grids can be economically more attractive in remote areas, they can have high up-front costs. This is particularly the case for renewable energy generation technologies compared with more conventional energy options such as diesel.

Financing has typically required some level of support from the local authority, NGOs, and international donors in order to pay for the capital equipment. Formal institutions such as banks or savings and credit associations are also common sources of financing. The 'Investments' section of Chapter 3 outlines types of investment financing such as venture capital, debt and equity financing, revolving loans, credit lines, and others. The same sources of financing apply to mini-grids but the most common of these are debt financing in combination with subsidies and grants.

Debt financing

Common debt financing for mini-grid delivery has been the leasing mechanism used to finance the sale of energy equipment and services, and is also used to finance energy service companies (ESCOs). Leasing gives the lessee use of the project equipment and products in return for regular payments to the financier who remains the legal owner until the loan is paid back fully. The ESCO model operates both at a macro level, where governments could be key players, and at a micro level involving end users, especially where the end users pay for the energy service provided to them by the ESCO (as depicted in Figure 4.2). In the ESCO model, governments and multilateral organizations (e.g. the GEF) provide financing. Examples of financiers for mini-grids include the African Development Bank, Asian Development Bank (ADB), and the European Bank for Reconstruction and Development. Local financiers include organizations ranging from banks to microfinance institutions, and savings and credit cooperatives. However, financing through commercial banks is still limited for mini-grids as the associated risks are considered to be high. Well-proven and scaled-up models are not yet in place, and debt financing takes place mostly if some sort of subsidy or grant exists simultaneously.

Subsidies and grants

Subsidies and grants have often been used to develop mini-grid energy delivery models, and then to continue to subsidize their energy tariffs, but evidence from an ESMAP report indicates they have not always been successful:

> There are many instances in which small [micro-hydro and local renewable energy] systems have been set up with little thought as to institutional support or maintenance, and many now sit in varying stages of disrepair. These systems often involved donor equipment that was provided to local communities through grants. The systems were typically installed without much local participation in design or system management. The result was that after a few years of operation, the systems often broke down, with little prospect of local communities being able to repair the equipment. (Barnes, 2005)

Part of the reason for the historical failure of mini-grid systems has been a lack of education, training, institutional support, and detailed planning for ongoing maintenance requirements. This section will outline some of these issues and the attempts that have been made to try to remedy previous failures.

Table 4.1 describes some of the most common subsidy schemes related to mini-grids and their advantages and disadvantages; note that these subsidies can often be translated to on-grid and off-grid contexts as well.

One of the more successful subsidizing schemes has been that of output-based aid (OBA), which has proved an incentive for private investment in several World Bank Group projects (Mumssen et al., 2010). The World Bank has used OBA where performance-based subsidies are provided for private-sector companies to pass on to consumers, and has been used to get mini-grid schemes off the ground; it could also be important for scaling up mini-grid energy delivery in the future. However it should be noted that subsidies can have both damaging and beneficial effects, as they can undermine incentives for consumers to make least-cost choices and for investors to develop alternative energy forms. They have also been shown to disproportionately benefit higher-income households, who tend to use more energy than poor households. Even when subsidies do benefit the poor, they can represent an unsustainable financial burden on the state, unless the funds can be managed in an innovative way. When subsidies are used for mini-grid systems, they should be used to enable access to energy, rather than decreasing the ongoing costs of the energy supply. For example, financing the capital costs of the mini-grid infrastructure, rather than interfering with the ongoing management and maintenance costs (Barnes et al., 2010).

The Bonny Utility Company in Nigeria is highlighted in Box 4.7 as an example of a subsidized mini-grid system. The example describes how a group of private utility companies helped a community to receive a sustainable energy supply

Table 4.1 Categories of subsidies

Subsidy scheme	Description	Advantages	Disadvantages
Investment-based	Capital subsidies from a major donor target all or part of the initial investment costs.	• Supports only economically viable projects • Relatively easy to implement	• Implies a cost reflective tariff • Doesn't guarantee to cover operation & maintenance costs, which may remain high
Connection-based	One-time subsidy granted according to the number of connections achieved, often through a PPP	• Incentive for investment in rural and remote areas. • Mobilizes entrepreneurship	• Risk of system being overstretched • Often lacks finance for operation & maintenance • Requires stable legal environment
Output-based	Subsidy of the electricity produced – most commonly a transition measure while system generates enough revenue to be profitable	• Strong incentive for private capital and entrepreneurship • Encourages PPPs • Can guarantee upkeep of operation & maintenance.	• Requires stable and ongoing refinancing • Must be compatible with private-sector objectives • Harder to implement
Lifeline rates and cross-subsidies	A lifeline rate subsidizes some or all of the cost for poorest users – cross-subsidies impose tax on richer users; often used to fund lifelines	• Effective for increasing access to poorer, rural communities • Potential source of revenue for rural users	• Can compromise financial viability of the project or local energy company • Can restrict itself by reducing consumption of larger users
Operation	Supports the operation of the system, but not the initial investment, bridging gap between affordability and cost recovery	• Helps to secure revenues for private actors • Incentive for mobilising private capital and entrepreneurship	• Little to no incentive to achieve economic sustainability • Discourages revenue generating activities • Hard to implement

Source: ARE, 2011

through a micro-grid scheme, initially as a gesture of goodwill and contribution to developing community relationships in the Niger Delta. The scheme has been reliant on ongoing subsidization, but the utility companies involved are now assessing whether such mini-grid schemes can become self-financing.

Development banks, in partnership with government departments, can be effective in developing appropriate regulatory frameworks for implementing and scaling up mini-grid systems, providing both tax incentives and subsidies to encourage investment. Box 4.6 details examples of scale-up success stories of micro-hydro mini-grids in Nepal, supported by the Agricultural Development Bank of Nepal.

Box 4.6 Micro-hydro supported by the Agricultural Development Bank of Nepal

Many developing countries have established development banks to help channel government or donor finance to priority sectors (OECD/IEA, 2011); an example is the Agricultural Development Bank of Nepal (ADBN), which has played a crucial role in channelling subsidies in the country through its initial adoption of micro-hydro technology in mini-grid energy delivery systems.

   The strategy was driven by NGOs and was a combination of building up the capability of the local turbine manufacturers, and development of a number of technical improvements (the electronic load controller and the use of electric motors as turbines). A significant part of the sector (turbines for milling grain) is financially self-sustaining, and receives no subsidized support. ADBN's role was eventually phased out when the Alternative Energy Promotion Centre (AEPC) was created in 1996. This is an autonomous body under the Development Committee Act, and is overseen by the Ministry of Science and Technology (MST). The mandate of AEPC is to promote renewable energy technologies to meet the needs of those living in rural areas of Nepal. Many donors, in particular Denmark's development cooperation (DANIDA), assisted with the strengthening of the AEPC and it is expected that mini-grids will be a key component of any new initiatives. In 2011, the Government of Nepal and development partners jointly agreed to support the formulation of a National Rural and Renewable Energy Programme (NRREP) (www.aepc.gov.np).

Source: Khennas and Barnett, 2000; authors' own research, 2013

Box 4.7 Bonny Utility Company – gas-to-power

The Bonny Utility Company (BUC) delivers electricity to the residents of Bonny Island in Rivers State in Nigeria's Niger Delta. This is a community development project from the Nigerian Liquefied Natural Gas company (NLNG), which provides free gas from its liquefied natural gas plant to generate electricity for local communities. The BUC started as an initiative by industry partners in NLNG (Shell, Total, ENI, and the Nigerian National Petroleum Company), together with representatives of the community (the Bonny Kingdom Development Committee) and the local government.

   BUC is a non-profit organization with at least 75% local employees. The local government has provided BUC with tax exemptions and other services, and has donated land. The Rivers State Government provides BUC with its operating licence and has seconded staff to the utility. The Bonny Kingdom Development Committee assists with community engagement and helps with issues such as employment and land rights. All third party contracts are competitively tendered, but preference is given to local suppliers over national and foreign suppliers. Industry partners are funding BUC and provide technical and managerial expertise. All stakeholders are represented on the board of directors.

   BUC provides reliable and affordable power to residential and commercial consumers of Bonny Island. Electricity is supplied free up to an agreed limit. Above that, every customer can obtain additional services at a subsidized rate via a pre-payment model. About 20 % of customers rely on the free tariff. The industry partners subsidize the BUC services; the aim is to evolve into a self-sustaining financing model. BUC has recently added water production, treatment, and distribution services to its portfolio.

   The payment for electricity was initially very controversial among local customers. However, once people began seeing the impact of a 95% reliable electricity service, they began to understand the importance of payment systems as a sustainable mechanism that guarantees the services and limits the wastage that results from free arrangements. Business has benefited because machines and computers can be on all day. Uninterrupted power has also enhanced public services: for example, Bonny General Hospital has been able to double the number of operations it carries out.

Source: Nigeria LNG, n.d.

Carbon financing

Carbon financing has also been a source of finance for mini-grid development although up until June 2011, only 15 clean development mechanism (CDM) projects (0.2 per cent of the total) had been designed to increase or improve energy access for households (OECD/IEA, 2011). However there is a general increase in the development of programmatic CDM projects, taking into consideration small-scale energy interventions, and geographical boundaries. According to EU legislation, carbon credits from projects post-2012 can be used in the EU Emissions Trading Scheme if projects are located in the least-developed countries (OECD/IEA, 2011). This provides an incentive for the private sector in these countries to use carbon financing. However the barriers of insufficient regulatory frameworks for a secure commercial environment in many of these countries have yet to be broken down, especially in Africa.

Social investments

It has become increasingly common for mini-grid systems to be supported through social investment. Innovative financing models are being developed to try to increase the role of venture capital in helping scale up successful mini-grid energy delivery models. An example of this is Husk Power Systems (HPS) in India, which has grown rapidly in just a few years and is summarized in Box 4.8. This energy delivery model supplies electricity to the villagers using biomass gasification technology, and uses a pay-for-use approach to raise revenues and supply electricity. This method of financing is particularly suitable to private enterprise-owned mini-grid systems, where investment costs are recovered through a profit-making tariff structure.

Box 4.8 Rice husk biomass power stations, India

HPS Power Systems is a for-profit business that installs mini-grid energy delivery systems in India that are powered by gasification of waste rice husk. The energy-rich gas produced is used as fuel for a generator engine, which then supplies electrical power through overhead cables to households within a 2km radius of the plant. Customers sign up to a flat-rate tariff (approximately $2.20 per month) and receive a 240v AC supply with enough power for two 15W CFL bulbs and a mobile phone charger. The company now receives roughly 80% of its income through electricity sales, with the remainder coming from Government subsidies. However, the initial capital was provided by grant-funding from several social investors, most prominently the Shell Foundation which provided $1.65 m for R&D, strategy and training.

HPS has now installed nearly 50 wholly-owned plants operating, with a further 17 under franchise partnerships. The enterprise plans to have nearly 2,000 plants installed and operating by the end of 2014, providing an affordable and sustainable electricity supply to 1 m households alongside thousands of local jobs and business opportunities.

Source: Ashden Awards, 2011

Microfinancing

Due to the high up-front costs of most rural electrification options and the low cash capacity of rural households, innovative small-scale financing must be provided. Micro credit, leasing and prepaid meters for fee-for-service provision seem to be the most promising options (Martinot et al., 2000). There are few rural credit institutions that have successfully reached rural populations and most of them are located in South-east Asia – the region where microfinance is more developed. Four publicly sponsored rural finance institutions that have been successful in providing credit to rural areas in this region and these are: Badan Kredit Kecamatan (BKK) and the Bank Rakyat Unit Desa (BUD) in Indonesia; the Bank of Agriculture and Agricultural Cooperative (BAAC) in Thailand; and Grameen Bank (GB) in Bangladesh. All four institutions have high outreach levels and have reached full self-sustainability. BKK and GB serve the very poor (i.e. outstanding loans of less than $100), while BAAC and BUD serve lower to middle-income farmers with average outstanding loans of $300 to $560 (Motta and Reiche, 2001).

Private-sector investment

There are a growing number of private-sector investors in mini-grid delivery systems; two prominent examples are the S^3IDF and E+Co. As described in Chapter 3, the S^3IDF is an organization that lends ongoing financial and technical support to small-scale enterprises in India, while E+Co used to operate on an international scale, providing loans and equity investments in clean energy enterprises to a wide range of developing countries. While it is important to note that E+Co has recently passed on its assets to another organization, Persistent Energy Partners, due to a restructuring, important examples can still be learnt from their experience, as highlighted in Box 4.9.

Corporate social responsibility (CSR) has also played a role in setting up mini-grid systems, such as Schneider Electric Nigeria, which recently donated a solar energy project to the community as part of the company's CSR commitment. In September 2011, Schneider Electric inaugurated 'Villasol', a solar-powered micro off-grid facility for decentralized rural electrification in Asore, Nigeria. This standardized solution consists of PV panels, a battery bank, and a battery-charging station as a decentralized communal recharge system. The facility is able to supply power to schools, entrepreneurial activities, health centres, and a water supply facility, as well as providing a basic electricity supply for up to 100 households. It has great potential to be replicated in other communities (Schneider Electric, 2012).

Capacity building

Due to the intrinsically local scale of mini-grid energy delivery models, the role of entrepreneurs, community leaders, community-based organizations,

---

Box 4.9 Sustainable investor – E+Co

E+Co was set up as one of the first and most influential 'social investors', investing in clean energy projects in 1994 with funds raised through collaboration with development agencies, NGOs, and foundations. It has since provided business development and technical support to over 1,000 firms in Asia, Africa, and South America. Additionally, over $15 m has been invested in 200 firms and this has resulted in the mobilization of over $183 m in capital and the delivery of clean energy to 4.3 million people. E+Co provided the following main services:

- Debt and equity investing
- Assistance and support for:
  - fundraising
  - microfinance revolving fund set-up
  - business development and strategy

The firm's focus was on finding innovative ways of addressing the needs of start-up clean energy firms in developing countries, particularly initial and growth capital, strategy advice, and business development, with a particular focus on initiatives that could supply energy services to the very poor. In 2007, $15.4 m capital was mobilized (88% loans and 12% equity), with funding from a range of multilateral and unilateral donors, investment banks, government bodies, and foundations (including Goldman Sachs, the ADB, US Aid Agency for International Development – USAID, and the German Agency for International Cooperation – GIZ). E+Co increasingly worked with agencies like the UN Environment Programme and national governments to create regulatory frameworks and environments favourable to small-scale clean energy investment. However, despite its success, it was restructured in 2013 (restricting all future investments but allowing the continued management of its existing investments), highlighting the difficulty of trying to bring private-sector investment to mini-grid energy systems.

Source: Aron et al., 2009

---

and cooperatives in coordinating the market actors and supporting systems on the ground is perhaps more important than it is for either on-grid or off-grid models. The UNDP *Bioenergy Primer* (Kartha and Larson, 2002) suggests that bioenergy mini-grid systems are likely to be most successful – indeed they may only be successful – if local coordinating institutions have a significant role in their design, implementation, and ongoing management.

Although these organizations are crucially important in ensuring the sustainability of mini-grid energy delivery systems, they often have limited capacity in a number of areas; this needs to be addressed if they are to flourish. For community-owned and managed schemes, capacity building is needed not only for the managers and operators, but also for the end users who are often overlooked. For example, demand management is particularly important for mini-grids, as they often have a quite limited supply of energy in comparison to on-grid systems. Very often, for small schemes, the electricity supply may be limited to a few hours or to a maximum power level. Thus, building up awareness for the consumers is very important. If mini-grids are operated as enterprises, the entrepreneurs and/or the committees managing the systems also need training in business skills, basic technical issues, and marketing.

It is also key for the management to learn about financing arrangements such as taking out loans, day-to-day financial control. Without these skills, very often mini-grid operators fail to sustain their project or enterprise.

As mini-grid energy delivery systems are designed to meet the energy needs of a community or group of users, evidence has shown that they are more likely to be successful if they start with a community-level needs assessment to determine priorities and potential benefits from the service (e.g. improved health, education, and agriculture). Such an assessment must look at the number of households and their various energy requirements, as well as local resources and potential investment opportunities, geographic location, and cultural and social setting. Energy-related decisions made at the household and community level depend on a household's perception of its needs and how the budget is allocated. However, these decisions are often not based on complete understanding of the health and livelihood benefits of energy technologies, for example the health problems associated with indoor air pollution (IAP) caused by less clean energy sources.

In some cases a community-wide solution (e.g. an information centre or solar-powered irrigation system) may have a greater overall impact on wellbeing than individual solutions such as solar home systems. The involvement of local end users in the design and implementation of assessments and analysis of results increases the likelihood of a development project being supported locally. This is highlighted in Box 4.10 which explains how a community biogas system in Pura, India was only supported when it met the needs of the community. The community need was for a clean and reliable water supply, as opposed to the need perceived by the local government – replacement of biomass with biogas.

Box 4.10 Biogas in India

Understanding the nature of energy demand, and being open to surprises is important. Early efforts by the Karnataka State Council to promote community biogas systems in the Indian village of Pura failed to gain the support of the communities, as they sought to make villagers substitute their usual fuel wood with biogas. As fuel wood is abundant, accessible, and free in Pura, people had no incentive to switch. Furthermore the volume of cow dung in the community was insufficient to provide a steady supply of cooking fuel.

However, the villagers subsequently revealed their desire for a clean and reliable water and lighting supply for homes. The community then established a system that produced biogas for fuelling a five-horsepower diesel generator. Electricity from the generator was supplied to households through a micro-grid and used to power a deep tube-well pump. There was also a radical change in the decision-making process, from community-based participation at the village level to top-down management by project administrators located over 100 km away in Bangalore.

Source: Reddy, 2004

Enabling environment

Unlike on-grid energy delivery, many national government actors tend to play a less dominant role in the delivery of mini-grid energy with much more localized market chains. However, this is starting to change, with local governments in a number of developing countries starting to play a much more active role in connecting populations to electricity through subsidized mini-grid schemes, particularly through low-carbon development (e.g. switching from diesel generators to renewable powered mini-grids). In Colombia, electricity supplied through traditional fossil fuel-based mini-grids is being subsidized by the Instituto de Planificación y Promoción de Soluciones Energéticas para las Zonas no Interconectadas (IPSE), a public institution that provides energy to rural communities not connected to the national grid. This is becoming increasingly common in many Latin American countries with high electrification rates and where access to electrification for rural or off-grid areas has become a public responsibility.

In addition, mini-grid energy delivery initiatives have often been initiated and driven by **civil society organizations**, with financial and capacity-building support being sourced from the public and private sector, indirectly influencing the enabling environment. However, policies and regulatory frameworks supporting mini-grids (such as the allowance of localized ownership, management and transmission of electricity, tax reductions or relief for energy equipment or raw materials) will continue to remain key enablers for mini-grids to be successful. Government will need to play an important role in defining the enabling environment for mini-grid energy delivery systems. The role of civil society is particularly important in building up information and awareness for the communities and entrepreneurs investing in mini-grids. In Peru, the government has finally started to recognize mini-grid energy systems as an integral part of the national electrification system, even making them eligible for subsidies. This has occurred after years of piloting and advocacy work by Soluciones Prácticas (Practical Action's Latin America office).

In Africa, the **public-sector** domination of the energy sector continues, with purely private sector-driven decentralized energy systems traditionally struggling. However this is gradually starting to change due to support from multilateral or bilateral donors. Many governments now have a Rural Electrification Agency or Rural Energy Agency (REA), to oversee the provision of energy to rural areas. There are current plans for energy for rural areas to be provided through mini-grid delivery systems, particularly in areas where the extension of the national grid has not yet happened. However it does need to be acknowledged that setting up such agencies does not actually result in any change taking place, and a lot of capacity building still needs to take place, from assessments of resources, through to the development of locally appropriate technology designs and implementation schedules, to ensure national and regional plans are actually delivered. An example of this took place in Nigeria, where international donors helped set up their renewable energy master plans and REAs, but then did not continue to support them. This has resulted in very little in the way of direct implementation of mini-grid energy delivery systems.

A number of national African **rural energy agencies** have recently been established and have managed to successfully create enabling environments allowing private-sector companies to participate in delivering energy to rural areas through hydro and solar-diesel hybrid mini-grid systems (Agbemabiese, 2008). These agencies include: Agence Malienne pour le Développement de l'Energie Domestique et de l'Electrification Rurale (AMADER) in Mali, Agence Sénégalaise d'Electrification Rurale (ASER) in Senegal, Rural Energy Agency in Tanzania, and the Rural Electrification Authority in Zambia. Similarly, in South Asia, governments, often with donor support, have created a 'reasonably-sized footprint in supporting modern energy, and have set up support agencies to help poor people move to cleaner options, including renewable energy, e.g. the Indian Renewable Energy Development Agency and the Alternative Energy Promotion Center in Nepal' (Morris et al., 2007). In Sri Lanka, the government, with the assistance of the World Bank and the GEF, implemented the Renewable Energy for Rural Economic Development (RERED) project to provide re-finance, grants, and technical assistance in order to promote mini-grid electricity services and encourage competition in the power sector (Morris et al., 2007). At the end of this project in December 2011, 59 projects were commissioned and mini-grid connections were installed with a total capacity of 148.8 MW. The total capacity addition to the national grid from all RERED-assisted projects (completed and under construction) will be 185.3 MW by the end of 2012. A further 175 community based mini-grid projects (173 village hydro and 2 biomass) were commissioned with a total of 1,769 kW and a further 110,575 solar home systems installed (off-grid) (RERED, 2012).

Governments also play a crucial role in promoting investments for mini-grid energy delivery systems, including investment incentives for local community and cooperative groups. In particular, governments are interested in renewable energy resources as these often have higher initial capital investment costs, but lower running costs. This type of investment is likely to be required even more in countries with poor infrastructure and weak business environments, as is the case in much of sub-Saharan Africa. In addition many governments in developing countries are not always transparent about their rural electrification plans, which can contain vague and over-ambitious mini-grid energy delivery plans. These plans are often counter-productive and actually reduce the confidence of investors and private-sector actors as they are not based on sound research. The false promise of grid connection is a common barrier to scaling up mini-grid systems as investors are unwilling to support community systems when they believe (often falsely) that a grid connection may come at a later stage.

In summary, much of the literature and practitioners' experience suggests that governments, at all levels, need to continue to help develop appropriate enabling environments for mini-grid energy delivery models. This will help to ensure significant involvement of private-sector organizations and investment financiers. For example, in the Philippines, renewable energy legislation was adopted in June 2009, creating strong interest through a range of tax breaks and tax incentives for mini-grid systems. In addition, foundations have been set up to develop renewable energy portfolio standards, including a feed-in tariff for renewable energy powered systems (ARE, 2011). Feed-in tariffs can be highly beneficial for initiating mini-grid businesses and can play a significant role in rural electrification delivery. If the regulations are catered to feed-in solutions, and the tariffs are commercially attractive, there is potential to generate much interest in establishing and managing mini-grids from the private sector, cooperatives or even local government (ARE, 2011).

CHAPTER 5
Off-grid energy delivery

Small-scale, off-grid technologies can provide energy to remote households and other isolated consumers that do not have a reliable electrical grid connection and do not want the ongoing costs of mini-grid systems. Rising fuel prices have further increased the economic attractiveness of off-grid technology options that rely on locally available fuels, often utilizing renewable energies. A significant portion of energy services will need to be delivered off-grid if universal access by 2030 is to be achieved, and there is a role for the private sector to provide the required financial resources and innovative delivery models. This chapter outlines the main issues surrounding off-grid energy delivery systems and how they can be effectively up-scaled to meet the enormous latent demand, presenting a number of examples that have been successful in some developing countries.

Keywords: fuels, mechanical power, off-grid, renewable energy, remote areas, up-scaling

Off-grid refers to the delivery of energy services and products to individual consumers (including households, businesses, public facilities, or community services) that are not connected to a national, regional, or community grid of any sort. Off-grid energy delivery systems use a wide range of energy resources to address energy service needs, and importantly do not only refer to the delivery of electrification. They also often provide a range of packaged and processed fuels for heating and cooking, as well as mechanical power for transport and agricultural processing. Off-grid systems can be complementary to services provided to grid-connected users, as well as to those currently unconnected.

For decades, grid extension, diesel-powered mini-grids, and mini-hydropower generators were, for the most part, the only electrification options available for rural communities. However, with the commercial maturation of a range of small-scale, off-grid (or stand-alone) products, an alternative has emerged for increasing access to energy for remote, isolated households. Such off-grid options are often best suited to those that do not receive a reliable electrical grid connection or are unlikely to in the near future, and do not want to be tied into the problems associated with mini-grid systems. Moreover, the dramatic rise in fuel prices has further increased the economic attractiveness of off-grid technology options that rely on locally available fuels. These off-grid technologies often utilize renewable energy-based technologies (e.g. power

http://dx.doi.org/10.3362/9781780447612.005

derived from sun, wind, and water) and, as a result, often do not have ongoing costs (World Bank, 2008), even if they do have higher up-front capital costs for the energy conversion equipment and appliances.

The sustainable development of off-grid energy systems requires more than just appropriate technologies. As these technologies can have a relatively high initial cost, they often require access to sustainable financing to ensure they are affordable. Development of appropriate infrastructure to support their distribution or installation is also needed. Other requirements include ensuring the services are effectively provided over the long run (particularly the provision of spare parts), and maintaining the systems (especially for remote, and often poorly educated, households). Effective prioritization and planning at the national, regional, or local level is also essential.

A number of off-grid energy delivery models have recently achieved a degree of success in some developing countries, and this will be explored in more detail in this chapter. However, in order to achieve universal access to a range of energy services, with a significant percentage coming from off-grid energy services (particularly for cooking and heating), there is still much progress to be made. The International Energy Agency (IEA) states that to achieve 'Energy for All' by 2030, approximately $7 bn will need to be invested annually in off-grid technologies, requiring a significant step up in financial resources and innovative delivery models, much of which will need to come from the private sector (OECD/IEA, 2011). As in chapters 3 and 4, Figure 5.1 outlines an example of a market map of an off-grid energy system, using the Azuri Technologies (formerly Eight19) solar lantern delivery model. This model includes the three key segments: the market chain actors, the most important supporting services for off-grid energy systems, and the relevant enabling environment. This chapter aims to outline the main issues surrounding off-grid energy delivery systems, and how they can be effectively up-scaled to meet the enormous latent demand.

Energy market chain

Off-grid energy resources

As with the on-grid and mini-grid systems discussed in the Chapters 3 and 4, any resource can be used to power off-grid systems. Fossil fuels are generally processed and packaged to be delivered in self-contained units, such as liquefied petroleum gas (LPG) cylinders, kerosene bottles and diesel canisters for household cooking and lighting appliances, and small business-scale diesel generators. Bioenergy resources are derived from agricultural waste products, (e.g. rice husks, coconut shells, cow dung, etc.), biofuels (e.g. ethanol and pure plant oil – PPO) and natural bio-resources (e.g. fuel wood from forests). Charcoal, whether produced from sustainably managed forests or harvested from natural forests, has traditionally been, and still continues to be, the most used off-grid cooking fuel in both urban and peri-urban areas. However using charcoal from natural forests often results in their gradual depletion, followed by other

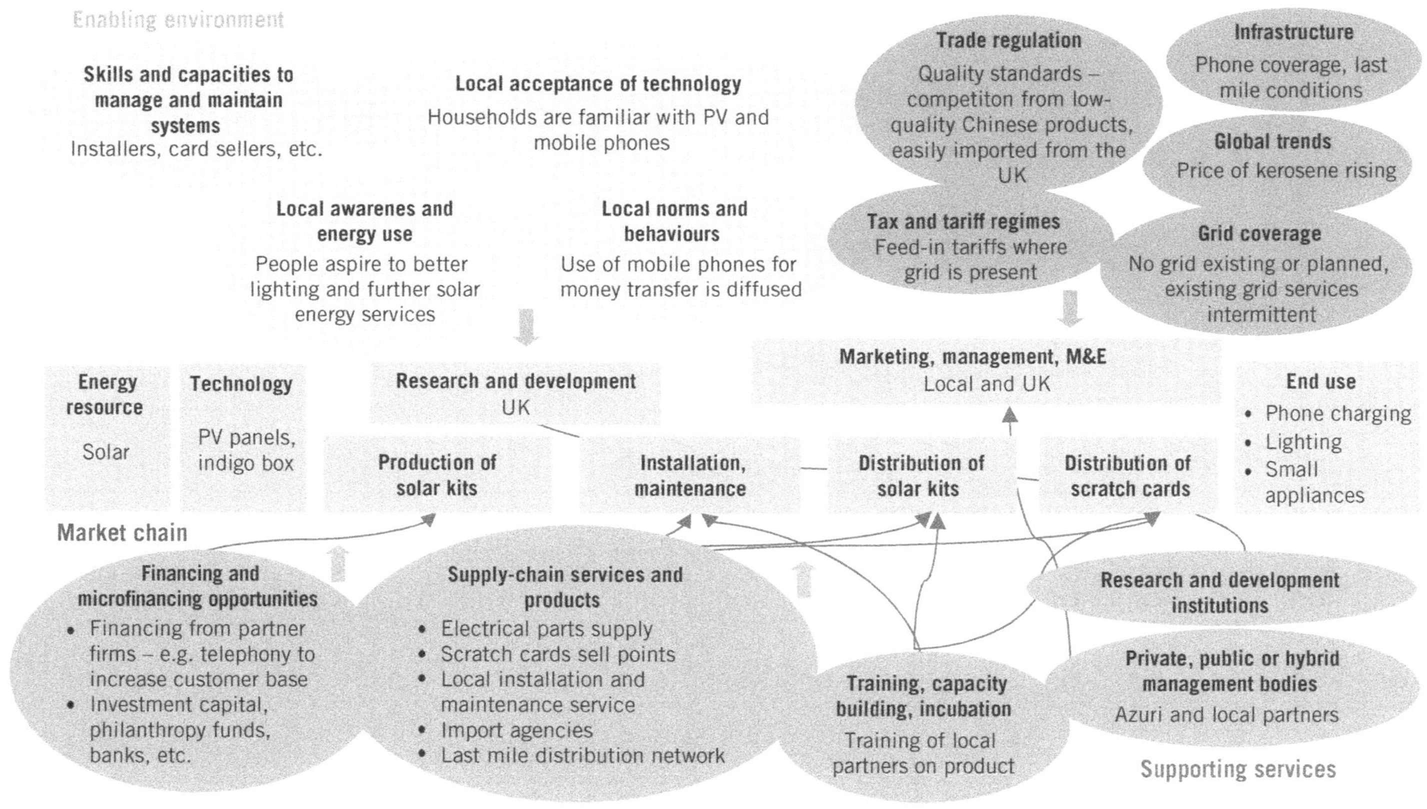

Enabling environment

Skills and capacities to manage and maintain systems
Installers, card sellers, etc.

Local acceptance of technology
Households are familiar with PV and mobile phones

Infrastructure
Phone coverage, last mile conditions

Global trends
Price of kerosene rising

Local awareness and energy use
People aspire to better lighting and further solar energy services

Local norms and behaviours
Use of mobile phones for money transfer is diffused

Grid coverage
No grid existing or planned, existing grid services intermittent

Trade regulation
Quality standards – competition from low-quality Chinese products, easily imported from the UK

Tax and tariff regimes
Feed-in tariffs where grid is present

End use
- Phone charging
- Lighting
- Small appliances

Energy resource	Technology	Research and development	Installation, maintenance	Marketing, management, M&E
Solar	PV panels, indigo box	UK		Local and UK

Market chain

Production of solar kits

Distribution of solar kits

Distribution of scratch cards

Research and development institutions

Private, public or hybrid management bodies
Azuri and local partners

Supporting services

Financing and microfinancing opportunities
- Financing from partner firms – e.g. telephony to increase customer base
- Investment capital, philanthropy funds, banks, etc.

Supply-chain services and products
- Electrical parts supply
- Scratch cards sell points
- Local installation and maintenance service
- Import agencies
- Last mile distribution network

Training, capacity building, incubation
Training of local partners on product

Figure 5.1 Off-grid market map of the Azuri Technologies solar lantern delivery system

negative environmental impacts such as soil degradation. Charcoal and wood fuel often constitute from 80–90 per cent of many countries' total cooking fuel usage, particularly in sub-Saharan Africa, and is delivered through off-grid energy systems. Pellets, normally made from agricultural waste or purposely grown crops, are also starting to be produced at scale in some countries, for example Kampala Jellitone Suppliers (KJL) in Uganda and Abellon CleanEnergy in India. Hybrid fuel solutions have also been developed (e.g. stoves, gasifiers, and biodigesters) that can use a range of types of biomass, as well as multipurpose products that simultaneously burn biomass for cooking and produce electricity. Examples of this type of technology include the BioLite and PowerPot stoves, as well as the SCORE stove, which is still in its research and development stage, although none have yet been extensively disseminated in developing countries.

As indicated previously, a significant percentage of off-grid energy needs are met by biomass. The delivery systems which surround the supply of biomass in many developed countries range from well-organized and well-regulated large-scale commercial networks for fuel wood collection, processing and distribution; to completely informal systems where individual households harvest and collect wood nearby their homes, outside any formal economic market. The legal and illegal trade of firewood for cooking and heating serves nearly half of the world's population, just in developing countries. The biomass is mostly sourced unsustainably and involves complex informal-sector market structures and supply chains, as highlighted in Box 5.1. The global production of wood charcoal has been estimated by the Food and Agriculture Organization (FAO) at 47 m tonnes in 2009, an increase of nine per cent since 2004. This development has been very strongly influenced by Africa, as it is still the region with by far the most significant production (63 per cent of global production). Charcoal production has grown in Africa

---

Box 5.1 Unsustainable use of biomass

The use of fuel wood has led to some localized deforestation, but the depletion of forest cover on a large scale has not been found to be attributable purely to demand for fuel wood. More often than not, fuel wood is gathered from roadsides and trees outside forests, rather than from natural forests. Clearing of land for agricultural development and timber are the main causes of deforestation in developing countries, but fuel wood can be a compounding factor. Studies at the regional level indicate that as much as two-thirds of fuel wood for cooking worldwide comes from non-forest sources such as agricultural land and roadsides. The scarcity of wood typically leads to greater use of agricultural residues and animal dung for cooking, which are then not available for use in the fields, resulting in reduced soil fertility and an increased propensity to soil erosion.

Charcoal, on the other hand, is usually produced from forest resources, with unsustainable production of charcoal in response to urban demand, particularly in sub-Saharan Africa. This places a significant strain on biomass resources. Charcoal production is often inefficient and can lead to localized deforestation and land degradation particularly around urban centres.

Source: OECD/IEA, 2006

by almost 30 per cent since 2004, extending its global dominance of usage (FAOSTAT-ForesSTAT, 2011), and the evidence seems to indicate this is unlikely to change in the near future.

Although the collection, processing, distribution, and sale of wood fuel, and the informal production of charcoal, are both largely unregulated in most developing countries, they still remain two of the most abundant and easily accessed energy resources available. This particularly applies to the very poor who are typically unable to access other cooking fuels, such as LPG or kerosene, due to their limited, or non-existent, incomes.

Biomass can be used quite efficiently and burned quite cleanly if the right technologies are available and used. It can also be supplied sustainably if the right management and processing systems are in place. However serious issues relating to the unsustainable sourcing of firewood and charcoal still need to be tackled if forests are to be left for future generations. A number of initiatives are being implemented to address some of these issues, including the UN Collaborative initiative on Reducing Emissions from Deforestation and Forest Degradation in Developing Countries (REDD+) and the Global Environment Facility (GEF)-funded projects. Box 5.2 illustrates a local initiative, which is focused on tackling the sustainable production of biomass, with the Bondo Farmers' Cooperative. The group promote a more sustainable approach to growing wood for charcoal production in western Kenya for use in off-grid cooking appliances, and which could be scaled up regionally.

Other off-grid renewable energy systems include pico-hydro (energy generation under 5 kW) and pico-wind (energy generation under 1 kW), as well as various forms of solar power (both thermal and photovoltaic – PV). Although off-grid pico-wind systems have been developed and piloted in a number of countries around the world, including Peru, Nepal, Nicaragua, and

---

Box 5.2 Sustainable production of wood fuel in Kenya

This initiative, which focuses on growing acacia trees for charcoal, was initiated by Youth to Youth Action Group (YYAG) and Thuiya Enterprises Ltd. in September 2002 in Kenya. Project beneficiaries include farmers, households, charcoal producers, transporters, wholesalers, retailers, community-based organizations, and Kenya Forestry Research Institute. YYAG sensitized and mobilized farmers interested in planting trees for charcoal on a commercial basis while Thuiya Enterprises Ltd. provided funding for the Charcoal Contract Farming Project. A six-year cycle was recommended to ensure maturity of trees.

Farmers were also given one beehive for every 500 trees planted. Harvesting of trees for charcoal started in 2008 and it emerged that six-year old acacia trees produce heavier charcoal than four-year old acacia trees. After seeing the final product and believing that acacia trees can be planted for charcoal, more farmers are now interested in planting and managing acacia trees for charcoal. The market focus and integrated design of the project is evolving into a self-sustaining initiative. This involves initiating co-benefit activities such as bee keeping that runs concurrently with the main project to keep some income coming in while the trees are growing. The number of beehives will be increased to at least one for every 100 acacia trees.

Source: Practical Action Consulting, 2009b

the Philippines, they are often very site-specific and often dependent on some form of donor funding. Questions still remain about their appropriateness for delivering energy at scale. Pico-hydro systems are also currently being developed in Asia, for example Laos, Nepal and Sri Lanka. However the two technologies, which have undergone the greatest uptake within off-grid energy systems are pico-solar (for PV systems with a power output of 1–10 watts), PV (particularly for solar lanterns and solar home systems) and biogas systems which will be explored in more detail in this chapter.

Energy market chain actors

As with the other energy delivery model systems, the off-grid energy system market chain starts with the energy resource, and includes all of the technologies (from the conversion equipment to the appliances), ending with payment from the end users. However, as both the conversion equipment and appliances need to be installed in each household, they need to be marketed and then sold through an extensive and targeted distribution system. During the initial design of off-grid energy systems, energy practitioners need to be aware of a range of critical issues when considering the technology choice and the energy market chain. They need to consider the initial fuel sourcing through to the design of the technologies to meet the needs of the end users, as well as the payment systems and distribution networks, to ensure effective installation and maintenance. As the very poor have limited disposable income, they are often discerning consumers with high expectations. They could be quick to dismiss systems which do not perform as expected or do not provide the required level of energy service. It is also important for the end users to be clear about what off-grid energy systems can deliver and what they cannot; this is illustrated by the sales of solar PV lanterns in East Africa, where end users have sometimes been very disappointed that the systems are only able to power LED lighting and mobile phones, and are not able to power larger appliances such as televisions and refrigerators.

The growing recognition of off-grid energy delivery models being able to supply energy services to the very poor and remote is demonstrated by a number of new, integrated development projects. An example of this is CleanStar Mozambique (see Box 5.3) where the delivery of efficient ethanol cook stoves to end users is just a small part of a whole energy system chain. The chain starts with the production of both food and fuel crops (including vegetables, cassava, and sugarcane), to ensure that fuel production does not compete with local food production; it then moves to processing in a centralized facility and then distribution of ethanol fuel to the urban market. This approach has been particularly successful as it has established an entirely new off-grid energy delivery model. The previous poorly performing wood fuel stoves have been replaced with more efficient, affordable, and clean-burning ethanol stoves, as well as providing livelihood benefits for the local population. The system is locally and appropriately designed to supply a higher quality and sustainable

---

Box 5.3 CleanStar Mozambique – a holistic approach to fuel supply chains

CleanStar Mozambique is a commercial business that focuses on integrating sustainable supplies of food and energy with environmental protection. The project is a joint venture between CleanStar Ventures, Novozymes (a Danish biotechnology company) and ICM (a leading ethanol technology company), with a Certified Emissions Reduction (CER) carbon-finance agreement with Merrill Lynch. One of the key drivers behind the CleanStar project is to replace the use of charcoal in Mozambique with a clean and sustainable supply of ethanol. CleanStar invests and operates using the growing urban demand for food and energy as an opportunity to build profitable rural initiatives in conjunction with ecosystem restoration and community development.

At the front end of the energy chain, the company works with smallholder farmers to develop agroforestry systems, providing technical expertise and capital equipment to grow a diverse range of crops (including cassava). These crops are then used for food supplies and the surplus is processed in a community-based, centralized infrastructure, built and operated by CleanStar. Here the food is packaged and ethanol is produced from the cassava. The products (packaged food, ethanol fuel, and ethanol cook stoves) are then commercialized and distributed in markets. These markets traditionally would have used imported food and charcoal for cooking.

CleanStar is looking at replicating its initiative in other countries, including Kenya.

Source: Bellanca, 2012e

---

cooking fuel, as well as providing additional benefits such as increased food production and the protection of forests through reducing reliance on wood fuel and charcoal.

The development of off-grid biogas systems is another example of an integrated approach that links energy, agriculture, and waste management, in order to provide both cooking and lighting services, as well as improve agricultural production (through the use of the bioslurry as a by-product).

Off-grid energy delivery models are also well suited to private-sector actors. High levels of technology innovation have been implemented by specialist enterprises, for example household energy technologies such as cooking, lighting, and powering radios and cell phones. The off-grid energy market is gradually being populated by a mixture of private and semi-private actors, ranging from entrepreneurs, to social business and NGOs, often with support from donors and government institutions. Often it is the combination of a number of organizations working together that has led to success in providing energy services to the poor.

A number of off-grid energy delivery models have been supported through the corporate social responsibility initiatives of large companies. These initiatives are mainly part of communication and public image strategies, but also, increasingly, can be a way to explore new markets. Recent projects by Philips, Unilever, Schneider Electric, Shell, Total, Enel, and Eni have demonstrated this. These companies are very interested in supporting small-scale off-grid energy delivery systems which can eventually be managed by local entrepreneurs, through the development of locally appropriate sustainable business models. Unilever has recently started to support the marketing of improved cook stoves in Kenya (together with its cooking products) in order to align its own business development with an important social cause – a win–win situation.

Many off-grid energy small businesses and social enterprises are aspiring to follow the dramatic growth trajectory of the mobile phones market in developing countries. This market grew way in excess of expectations, and showed that even poor households (in both urban and remote rural areas), are able to purchase technologies when they are provided with services that are required (World Bank, 2012). This potential for growth is particularly the case for solar products such as the Sun King Pro. This product can charge mobile phones as well as lights, so is complementary to mobile use, particularly in areas with no grid or just mini-grid coverage. Companies such as Azuri Technologies, ToughStuff, Greenlight Planet, Barefoot Power, and d.light produce a range of appliances such as off-grid solar lanterns and solar home systems which provide lighting and mobile phone charging. Companies such as Envirofit, Stovetec, Ecozoom, and Ugastove produce and distribute improved cook stoves, but are not purely profit driven and have social targets in addition to financial ones.

However, although social enterprises have recently achieved some success in delivering off-grid energy systems (mainly for lighting and cooking), they have struggled to reach the very poor and those in remote rural areas. In contrast, certain government- and donor-driven schemes, such as the Project for Renewable Energy in Rural Markets (PERMER) rural off-grid electrification project in Argentina (detailed in Box 5.4), have been able to successfully deliver off-grid energy services to households in rural areas traditionally not reached by private-sector initiatives. It should be noted that such projects are often implemented in collaboration with private-sector organizations despite being initiated by donors. Such programmes are not based on a sustainable market

---

Box 5.4 PERMER rural off-grid electrification, Argentina

The Project for Renewable Energy in Rural Markets (PERMER) was introduced in Argentina in 1999, and focused on increasing electricity connections in rural areas through the provision of energy services powered by a range of renewable technologies (including off-grid household appliances). The primary actor in initiating the programme was the national government, which formed a public–private partnership with enterprises and played a central role in driving the implementation of the project, organising supporting services and creating the enabling environment. Funding was secured from several multilateral organizations (including the World Bank), and contracts to run and maintain the installations were signed with public, private, and cooperative concessionaires. Provincial governments coordinated with concessionaires and local populations to decide where off-grid energy systems were to be installed and at what tariff rate. The programme was predominantly top-down and focused on increasing energy access rather than making profits; this has been a significant barrier to expansion, as many private-sector concessionaires have been unwilling to serve unprofitable rural areas.

Source: Best, 2011

system, but rather on output-based aid; thus, it is often the case that off-grid initiatives, need to involve joint partnerships between the public sector, the private sector, and local communities, to really reach the very poor and those in rural areas.

Energy technologies and their design

While grid-supplied energy is almost always in the form of electrical power, off-grid energy delivery has a much broader range of contexts as a variety of technologies are available. The main technologies include: domestic heating and cooking equipment, and fuel products (stoves using charcoal, fuel wood, pellets, and briquettes); solar water heaters and boilers; domestic electricity-producing systems (including small generators which run on diesel, biodiesel and/or PPO, solar home systems, and stand-alone wind turbines); self-contained solar lamps (consisting of lamps with bulbs, solar panels, wiring, and battery packs); solar dryers, fridges, and phone chargers; and a range of mechanical power systems (e.g. water treadle pumps and individual grain-milling systems).

As with mini-grid solutions, off-grid energy technologies can be more specifically adapted to the end users than on-grid, which tends to be based on the 'one size fits all' model (a direct electrical connection via an overhead low voltage cable). In the case of remote communities, off-grid energy technologies often represent the only available technology choices, but contrary to mini-grid, do not require significant levels of in-built infrastructure or management models. However, the technologies currently on the market, and which are affordable to poorer users, are often unable to fully match the level of energy supplies that can be delivered through on-grid or mini-grid systems. Off-grid solutions tend to address individual energy needs separately, with household using a range of technologies and fuels, including candles, kerosene lamps, and solar PV systems for lighting. Electricity is rarely used for cooking and space heating, and biomass is rarely used to produce electricity within off-grid systems. As previously mentioned, biogas systems can represent an exception, as they can sustainably provide clean and efficient energy for cooking and heating, as well as for lighting (and the larger systems can even be used to produce electricity).

Up-front capital costs

The up-front cost of off-grid energy delivery systems is often a significant barrier to their uptake at scale, since the end user is generally required to pay for the entire technology system in advance. This is especially true in the case of cooking, where households are required to pay for the stoves and fuel canisters at the beginning, such as for wood, charcoal, LPG, biogas and kerosene stoves and containers, in addition to the ongoing energy resource costs of the fuels. For renewable energy systems, such as solar, hydro and wind systems, the up-front capital costs are still a barrier, but can be offset by the lack of ongoing fuel costs (this is also the same for collected fuel wood).

An example of this is the solar home system, where the end user is required to pay the entire cost of the energy conversion equipment (the solar panel, wiring, power regulator etc.), as well as the appliances (batteries, light bulbs and associated wiring) up-front. In the case of liquid petroleum gas (LPG), the cost of the stove, the cylinder, and the compressed gas within the cylinder, is together often too high for many low-income households. The refilling of an entire LPG bottle at once, which represents the fuel required for several months of cooking for a typical household, is often way above a typical households savings, even though the per unit cost of the gas might be within their reach. Poor households often have very low levels of savings and tend to buy products, including fuels, in very small units, such as a single candle or a small tin of kerosene. Bulk investments for off-grid energy systems often do not match the purchasing patterns of poorer families, precluding them from adopting cleaner alternatives to biomass, and locking them into poverty. In comparison, for on-grid energy, the infrastructure has generally been installed over long time periods and paid for by all citizens through taxes, low interest loans, or grants.

To try to overcome this barrier, several new innovative financial and payment models have been developed for off-grid energy systems. The LPG industry, faced with the problem of investment in cylinders and gas refills, has begun experimenting by introducing smaller bottles or 'top-ups' in smaller quantities, rather than refilling the entire container, and even giving the containers away for free to try to overcome the entry cost barrier. Such approaches effectively shift the business model from paying for all the equipment, to just paying for the service (this is generally the case for on-grid consumers, where the energy utilities pay for and own the energy delivery infrastructure). Box 5.5 provides an example of a piloted LPG initiative in Kenya, which demonstrates the opportunities and challenges of such a top-up system.

Box 5.5 LPG affordability – piloting top-up options

An example of an LPG delivery model called PIMA Gas is being piloted in Nairobi, with International Finance Corporation and World Bank funding. It features a 1 kg cylinder with a special dispenser that allows people to buy small amounts of LPG at 50–100 Kenyan shillings (around US$0.65–1.25), purchasing LPG the way they purchase kerosene or charcoal. The bottle is provided by the scheme free of charge. This is in the pilot stage, and there are still concerns around security of the transfer process during the top-up operation or during the transport to the top-up station. Currently the top-up stations are controlled by the utility company, but it is hoped that eventually small shop owners will be trained to safely provide the service so that they can hold on to their kerosene-buying customer base.

Source: Bellanca, 2012c

In another example from Uganda, a social enterprise, Foundation Rural Energy Services (FRES), is installing solar home systems in rural households and then selling the energy services to the users. FRES continues to own the systems, allowing households to obtain the energy services at competitive prices without having to pay for the infrastructure systems themselves. Other business models are being developed where the cost of the infrastructure is included within the cost of the fuel; the companies delivering the off-grid energy services offer the equipment to the users for free, or at a subsidized rate, and make their profits through the ongoing sale of the fuel. Box 5.6 highlights an example of an off-grid energy business model from India which centres around the use of carbon credits to reduce the costs of processed cooking fuel (in this case biomass briquettes), rather than users attempting to subsidize the cooking stoves themselves. The same model could be applied to other biofuels such as ethanol, pellets, and plant oil.

Similarly there are examples in the solar PV sector of companies, such as Azuri Technologies, which have developed innovative methods of allowing

---

Box 5.6 Biomass briquettes from forest waste, India

Management: private sector
Distribution: private
Scale: local (small scale)
Financing: private investment, regional subsidies, and carbon financing
Implementing organization: Rural Renewable Urja Solutions Pvt. Ltd

Rural Renewable Urja Solutions Pvt. Ltd (RRUSPL) is a private limited company located in Uttarakhand in northern India which manufactures and supplies biomass briquettes using pine needles and other forest residues and agricultural waste. Millions of tonnes of biomass are generated from forest residues, particularly pine needles (as these can cause a lot of damage to the environment causing fires and preventing water and light from reaching the soil if not removed from the ground).

RRUSPL initially involved 50 villages in the project and helped form self-help groups of 8–10 members (mainly women) to collect biomass from the forest. A cluster operator coordinated with the villages to assemble the biomass and produce the briquettes, and then supplied them directly to industries. They are used for various purposes, including in brick kilns and industrial boilers, and are used by restaurants, schools that run mid-day meal programs, ashrams, cafeterias, and school hostels. These target markets had primarily been using coal or LPG for their energy requirements.

The composition of the briquettes is 60% dry pine needles, 30% sawdust and 10% other agricultural waste (lantana, cow dung, and sugar mud). The stoves, which are specifically designed to burn the briquettes in a clean and energy-efficient manner, are provided to the end users for free, while profits are generated by selling the briquettes, gasifier *chulhas* (improved cook stoves), and by selling the carbon credits generated from the reduction in carbon emissions. The Climate Protection Partnership (My Climate), a Switzerland-based agency, has entered into an agreement with RRUSPL to buy the carbon credits generated by the project. There is scope for setting up five such units in the pine and lantana regions of various hill states in India. The current unit can also be expanded to produce a total of 10,000 tonnes of biomass briquettes per annum from its current capacity of 4,000 tonnes. The net result will be the prevention of 15,000 tonnes of greenhouse gas emissions.

Source: Winrock International India, 2010

the poor to access their services, such as by leasing their equipment through payments for the energy services they provide rather than the full cost of the equipment (mainly electrical lighting), following the model developed by the mobile phone industry. Azuri Technologies, a UK-based company, offers a 'pay-as-you-go' service, whereby a solar home system is installed for free or with a small deposit and consumers purchase scratch cards to activate it. After a period of 18 months, assuming a level of fairly constant use, the equipment will have paid for itself, and the customer can then choose whether to upgrade their system or buy it from the company. A simple system that powers two lights and a mobile phone charger is the first step, but can be gradually upgraded until a full set of household appliances can be powered. It has been calculated that, overall, users spend approximately half the money they would have on kerosene lighting over the same period, but have a complete working system by the end (Bellanca, 2012a).

Affordability

Traditional donor-driven implementation models are gradually being superseded by innovative, entrepreneur-led market initiatives which are profit driven and therefore more sustainable. Where off-grid energy delivery models aim to create sustainable markets for energy conversion equipment and appliances (as is usually the case with improved cook stoves and solar PV lighting systems), the existing income levels and spending priorities of the end users need to be taken into account to ensure their success. In general, very poor people prioritize price, while the less poor are able to pay slightly more for other technological features, such as quality and appearance (pers. comm. with Ned Tozun, 2012). It is believed that rising rural incomes in China may have been a significant factor behind the large-scale adoption of improved stoves (Sinton et al., 2004; Qiu and Gu, 1996), and in Africa, studies have shown that middle-income families adopt improved stoves more quickly than poor families (Barnes et al., 1994).

This suggests that purely market-focused improved cook stove initiatives would probably achieve greater penetration if they were aimed at areas with rising incomes, rather than the most deprived areas; conversely the very poor may not be adequately reached through purely private-sector approaches. The poor are generally more easily put off by high up-front costs, even if the technologies allow them to make long-term savings or gain benefits such as reduced indoor air pollution (usually more difficult for companies to demonstrate and users to appreciate) (Tsephel et al., 2009). In reality the cultural and socio-economic status of poor households can even make them reliant on indoor smoke for eliminating insects from their homes, or for preserving meat. These practices are often perceived as being more important in the short term, with the health risks of indoor air pollution often not fully understood within improved cook stove programmes (Boiling Point, 2010).

These issues of affordability can often be overcome through innovative credit schemes designed to spread the cost of the cooking appliance, or through some form of subsidization (e.g. carbon finance), to reduce the cost of the application. It is also important to understand that a product or service needs to be affordable in relation to local income and overall household expenditure, but it also needs to have a 'perceived value' that is greater than or equal to its 'perceived cost'. Overcoming resistance and scepticism about new products and services; convincing people of their benefits; and offering alternatives that meet their needs (e.g. encouraging households to replace smoky appliances with improved cook stoves or to use smoke hoods), are all vitally important for the acceptance and ultimate purchase of off-grid energy systems. The end users will purchase new off-grid energy technologies if their perceived value is greater than their cost (Bruce et al., 2000). One important difference between energy-saving cook stoves and electricity products is that the perceived value of electricity is often much higher than the stoves. Cook stoves often suffer major challenges to adoption, as they do not appear to have an immediate perceived value and usually require a change in the user's behaviour (Rai and McDonald, 2009). These issues of perceived value can be greatly affected by the income of the user, the cost of the fuel (a technology that can reduce this cost will be more highly valued), and the social status of the end user (Barnes, 2005; Cecelski, 2004; and Agbaje, 2009).

However it is also important to understand that the very poor may simply just not be able to afford the full cost of off-grid energy services, and in such cases some element of subsidy will be required to help them achieve Total Energy Access to a range of modern energy services.

Supporting services

In the following section, several of the most important supporting services for off-grid energy delivery are outlined, in particular the appropriate financial services and relevant capacity-building organizations and activities. These supporting services ensure that off-grid energy systems are designed and delivered to a high quality and appropriate design, so as to be effectively taken up by poor people living in a wide variety of social, cultural, and geographical settings in developing countries.

Financial services

As already mentioned, obtaining external financing to help end users meet the up-front capital costs on a range of off-grid energy delivery systems or products is often one of the main challenges to their scaling up (ARE, 2011). Large initiatives comprising small-scale, localized energy delivery systems are generally far less attractive to major investors who are used to on-grid energy delivery models where they receive significant returns on their investments through tariff payments in a low risk environment. Lighting Africa, an International Finance Corporation (IFC) initiative, estimates that 600 million people without grid connection in sub-Saharan Africa spend a significant percentage of their household incomes on fuel-based lighting, amounting to almost $10 bn per year, and about $25–38 bn globally. This income could be allocated to off-grid lighting technologies that can provide energy services to the poor such as solar PV lanterns. However, companies are deterred from investing in the sector due to the lack of adequate market information on both lighting and non-lighting products (Lighting Africa, n.d.).

This lack of market information is one of the main reasons why off-grid energy delivery programmes have traditionally been donor-driven (based on direct financial aid from governments, multilateral organizations, NGOs, or a mix of several organizations), rather than significant direct investment by the private sector. An example of this approach is highlighted in Box 5.9 which focuses on the gradual commercialization of the improved cook stove market in Sri Lanka, which developed over several decades through a variety of donor projects (Practical Action Consulting, 2009a; Rai and McDonald, 2009), each driven by either a state actor, NGO or a combination of both. Although the improved cook stove market in Sri Lanka is now purely private-sector driven, it was only able to develop through a mixture of financial input from both public-sector and private-sector funds.

However, the availability of financial services specifically designed and tailored for off-grid energy markets is slowly increasing, although a purely market-driven approach, without any form of subsidy (either direct or indirect), is still not very common. An extensive study published in 2007 estimates that the total base of the pyramid (BoP) household energy market in Asia, Africa, Eastern Europe, Latin America, and the Caribbean is approximately $433 bn, representing the spending of 3.96 billion people (Hammond et al., 2007). Note that the study focused on a consumer segment where households had an annual income of less than $3,000, with average household spending on energy being about nine per cent of their total income. Households on the lowest rung of the economic scale, defined as BoP 500, spend around $0.40 a day on energy, rising to $1 a day for those in the next segment up, BoP 1,500. The study also highlighted the differences in expenditure between rural and urban households; in Brazil rural households spend on average $102 on electricity while urban households spend $397. This study has led to a number of new initiatives designed to help develop a range of financial services specifically for BoP energy markets for off-grid energy technologies, particularly microfinancing. For example, the UN's capital investment agency's (UNCDF's) new clean energy financing programme, CleanStart, is due to be piloted in six countries in Asia and Africa. In addition, the EU Energy Facility SEMA project is building the capacity of off-grid energy companies and microfinancing companies in Uganda, Kenya, and Tanzania (SEMA, n.d.).

Grants and social business

Grants and subsidies have often been used to support off-grid energy delivery models that specifically target low-income markets to get off the ground, comprising both financial services as well as technical assistance. One of the main reasons for providing such grants to help the poor, and/or rural populations obtain a range of off-grid energy services, has been the improvement of livelihoods, including positive returns on health, education, and income. However, multilateral and bilateral agencies have started to provide grants and low-interest loans specifically focused on trying to leverage private-sector actors to stimulate the BoP energy markets; with the expectation that they will eventually generate good returns. Such grants and subsidies can play a vital role in stimulating the market to successfully deliver a range of off-grid energy services, in rural and peri-urban markets (as highlighted in the India case study in Box 5.7).

Box 5.7 Successful implementation of subsidies in India

Sometimes the removal of subsidies can improve the poor's access to energy, as in this example from India. Even when subsidies do benefit the poor, they may represent an unsustainable financial burden on the state, and market liberalization has been used as an alternative strategy.

In Hyderabad, India, only the richest 10% of households were using LPG in 1980, with middle-class households typically using kerosene because they could not obtain LPG (a more efficient and cleaner burning fuel). There was no kerosene available for the poor because the limited amounts available for public distribution were bought by middle-class households. As a result, the poor in urban areas had to pay for fuel wood, which was even more expensive than kerosene.

When the Indian Government liberalised the energy markets in 1991, and relaxed restrictions on the production and import of LPG, more middle-class households switched to LPG, freeing up supplies of kerosene for the poor. This led to over 60% of households in the cities starting to regularly use LPG.

Source: Barnes et al., 2005

Funds have been provided through overseas development aid (ODA) to set up loans which are then used to stimulate the market, as was the case with the Renewable Energy for Rural Economic Development (RERED) initiative in Sri Lanka. This initiative helped foster local private energy service installers, through the provision of subsidies to encourage local banks to provide access to credit for the end users (World Bank, n.d.). Many socially responsible companies who supply off-grid energy services have been able to take advantage of the support of donors, NGOs, and the general public to co-finance their initiatives, through public–private partnerships. Such models highlight that attempting to supply all the energy service needs of the poor is often not feasible on a purely commercial basis (Bellanca and Wilson, 2012).

Box 5.8 highlights one of the most widely known improved cook stove programmes in Sri Lanka, which involved private-sector actors together with a combination of GEF funds, bank loans, and credits. The credits were offered to bank dealers to lend directly to customers so that they could purchase the clean cook stoves (Practical Action Consulting, 2009a). The goal was to expand the stove market in Sri Lanka by making credit and partial subsidies available to the poor.

Box 5.8 Anagi household stove, Sri Lanka

Management: NGO, government, and private sector
Distribution: commodity market
Scale: national (small to large scale)
Financing: donor
Implementing organization: international cooperation, government, and private sector

The *Anagi* is a two-pot-hole pottery improved stove, originally developed and disseminated in Sri Lanka, and has been one of the most successful improved cook stoves in Asia. The programme had three distinct phases: stove design and piloting; small-scale dissemination; and large-scale dissemination and full commercialization. It was initiated through a joint project of Practical Action and the Ceylon Electricity Board in the mid-1980s, and between 1987 and 1990 about 80,000 units were sold. This initial success encouraged Practical Action in 1990 to embark on a joint project with the Integrated Development Association (IDEA), a local NGO. Potters and artisans were trained to produce the stove and market it. Advertising campaigns were also run through conventional media outlets.

Since 1991, an estimated 3 million *Anagi* stoves have been commercially produced and marketed throughout the country with a commercial network driving the distribution of approximately 300,000 stoves annually (Rai and McDonald, 2009). The market chain is well established. Approximately 185 trained potters spread over 14 districts manufacture the *Anagi* stove. Distributors and wholesale buyers visit production centres to buy the stoves in bulk, thereby guaranteeing the producers a regular and reliable market. The stoves are then distributed to retail shops throughout the country. Small producers often sell their products locally.

A number of factors led to the success of the *Anagi* stove. Technical development and considerable local field testing meant that the *Anagi* suited both local manufacture by local potters and also local cooking practices. Improved quality assurance on the product gave greater consumer confidence, and national media promotion plus local promotion by producers and retailers led to rapid wide-scale dissemination of the stove. Long-standing donor support for over two decades was crucial to trialling the different technologies and assisting in the commercialization.

Source: Practical Action Consulting, 2009a; Rai and McDonald, 2009; Amarasekara and Karunatissa, n.d.

Such off-grid energy programmes appear to work better in countries that already have a thriving entrepreneurial market, as well as extensive dealership networks and marketing channels. In East Africa, output-based financing for household solar PV systems has been promoted by the World Bank. One programme in Tanzania, implemented through the Tanzania Energy Development and Access Project (TEDAP), aims to reach remote rural households, in a country where only 10 per cent of all households, and less than 2 per cent of rural households, have access to electricity. As the national grid cannot be economically expanded into many rural areas in Tanzania, a mechanism to try to incentivize and fund off-grid household solar PV systems was set up. Performance-based grants were co-financed by the International Development Association ($7.6 m) and GEF ($2.1 m), and managed by the Rural Energy Agency (REA). These were then made available to private enterprises, cooperatives, or NGOs for the installation of solar PV systems in rural households. The incentives were based on the actual number of new households supplied with electricity from such systems, although they were only partly paid for up-front, reducing the amount of pre-financing required. The overall payment structure was arranged between REA and the developer with a grant agreement. Upon mobilization of the target households, 40 per cent of the costs were provided up-front to the developer; 40 per cent were paid after the delivery of the equipment goods to the site; and the final 20 per cent were paid after approval from the end user and REA. The subsidies were provided at different levels with the highest levels for the smaller systems, gradually being phased out for larger systems, to ensure the poor benefited the most.

Such performance-based output aid financing is a good example of the multiple uses of donor and private funds through public–private partnerships. Similar initiatives have also been carried out using slightly different mechanisms in Uganda, and several other countries. For example, providing solar PV off-grid energy technologies to rural areas through targeted financial subsidies and grants to stimulate companies which are willing to operate in areas further away from their urban bases.

Debt finance

Debt financing. This is a common mechanism which allows financial institutions or capital investors to invest in services, such as off-grid energy delivery, in order to earn interest and a return on capital through the end-user payments. For off-grid energy technologies it is common for all the capital equipment to be purchased up-front through the use of debt financing, and the cost to be recovered through regular instalments paid for by the end user over several years. In Bangladesh, Grameen Shakti has been using this model to enable households to obtain solar PV and biogas systems, as highlighted in Box 5.12. Two common debt-financing models, consumer credit and dealer credit, are outlined in Figures 5.2 and 5.3 (GVEP International, 2010).

Consumer credit model. In this model, two financial players are involved (commercial financiers or donors and/or NGOs) in providing loans or loan guarantees to local financiers. These financiers then advance credit to end users to purchase goods from energy enterprises. In East Africa, local savings and credit cooperatives (SACCOs) have started taking out loans from the national banks in order to lend them on to their members, at slightly higher interest rates, for them to purchase off-grid energy technologies (particularly solar home systems). The consumers then purchase the equipment (solar panels or biogas systems) directly from the energy technology enterprise or supplier, for example, Barefoot Power.

Commercial financier/donor → Local financier (eg. MFI/SACCO) → Consumer ← Supplier (enterprise)

Figure 5.2 Consumer credit model
Source: GVEP International, 2010

Dealer credit model. In this model, financial institutions provide credit to the energy technology suppliers or energy entrepreneurs who then sell the products on to the consumers together with credit in the form of a loan (e.g. Solar Now in Uganda). The financial institutions are not responsible for the credit to the consumers, with the energy companies taking on a much higher level of risk (often in an area where they lack expertise). Solar Now is hoping to find a financial institution that can take over the provision and management of the loans.

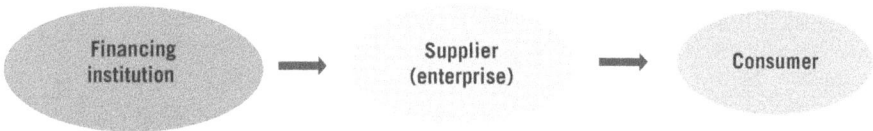

Financing institution → Supplier (enterprise) → Consumer

Figure 5.3 Dealer credit model
Source: GVEP International, 2010

These are only two of the financing models for off-grid energy product and services, and new variations are emerging all the time; it must, however, be noted that the effective implementation of such models is still in its infancy. A recent study in East Africa showed that financing provided by financial institutions specifically for energy technologies tended to be extremely low, unless a specific programme was pushing for it. Many banks and microfinance institutions (MFIs) do not have specific energy portfolios but lend as part of a range of working capital and asset finance products, for example home loans which can sometimes be opened out to include a solar home system or improved cook stove. The most common energy products financed, in both rural and urban areas, were found to be solar lanterns or solar home systems, generally for domestic use. The most common barriers to scaling up lending for off-grid energy technologies included a lack of available financing at all levels; inadequate capacity within the lending institutions; and a lack of awareness of the technologies within the financial institutions as well as by the consumers (Kariuki and Rai, 2010).

While the role of debt financing for off-grid energy technologies is believed to be crucial for stimulating the energy market chain to ensure appropriate quality products reach the end users in order to meet Total Energy Access, it must be noted that market intelligence is still quite weak, especially for the rural poor. Another example is a treadle water pump initiative in Bangladesh implemented by International Development Enterprises (IDE), where different quality standards for the different prices of products had been developed. While the most expensive model had an expected life of at least seven years, implementers found that the cheapest model, with only a two-year predicted lifespan, immediately captured about 50 per cent of the treadle pump market when it was rolled out. End users were purely basing their choice on the price (Polak, 2003). This choice was made because poor smallholder farmers were typically short of cash and so bought the cheapest pump with a view to upgrading to a better model once increased income had been generated. Debt financing was arranged through a number of financial institutions and the private sector was involved in creating a vibrant market chain, enabling farmers to afford the better quality products at the beginning. Farmers were able to obtain a better quality product to significantly increase their production, and pay back the loans in a short period, resulting in a win-win outcome. When such market intelligence is available and can be regularly fed into the market chain, in coordination with the design of effective debt financing, it can result in increased energy access in rural poor areas of developing countries which can mean significant livelihoods benefits.

Carbon finance

Carbon financing for off-grid energy delivery models is another emerging financial service, increasing access to off-grid energy services through the subsidization of the energy technologies or their delivery mechanism. The design of carbon financing varies greatly, with enterprises that use carbon finance principally to develop their business; those that attempt to stimulate the overall market, or marginally decrease the technology costs to the end users; as well as those that completely rely on carbon credits for the market to exist at all. An example is C-Quest, a fully commercial enterprise which requires a minimal contribution from their end users in order to install improved built-in cooking stoves, while using carbon credits to cover the remaining balance. Practical Action and Carbon Clear have recently received Gold Standard carbon accreditation for their clean cook stove project in Darfur, Sudan, helping poor families use carbon credits to replace their inefficient and unhealthy wood and charcoal stoves with clean burning LPG stoves (Gold Standard, n.d.).

Solar products are also eligible to receive carbon revenues. Some companies already have projects registered, although they are not expected to produce very large levels of subsidies due to their relatively low levels of carbon abatement. D.light design has already received approval from the UN Framework Convention on Climate Change (UNFCCC) for a ground-breaking carbon offset project in *Uttar Pradesh* and *Bihar* in India. The project innovatively tracks the reduction in carbon emissions that result from the replacement of kerosene lanterns with D.light solar lamps. It is the first of its kind to receive UNFCCC approval but it is likely that others will soon follow (CDM Executive Board, 2009).

Although carbon revenues can support the financing of off-grid systems, their high transaction costs are often barriers to local producers accessing them. In response, a number of initiatives have been developed to try to overcome these barriers, such as Nexus-Carbon for Development, which brings together an innovative alliance of social ventures (including non-profits, NGOs, and eco-businesses), to try to help scale up best practice for carbon financing of off-grid energy access. This contributes to climate change mitigation as well as poverty alleviation (Nexus-Carbon for Development, n.d.). Another example, from Nepal, is described in Box 5.9, and outlines how carbon finance has been provided as part of a whole programme to help reduce transaction costs.

Box 5.9 Carbon financing, Nepal

Management: household
Distribution: stand-alone
Scope: local (small scale)
Financing: multilateral donor and NGO
Implementing organization: NGO and civil society

The Biogas Support Program has been running in Nepal since the 1980s, and is a collaboration between a range of organizations from government, business, civil society, multilateral donors, and the Netherlands Development Organization (SNV). Since 2007, the Global Partnership for Output-based Aid (GPOBA) has provided annual funding of $5 m to subsidize the implementation of 37,000 household biogas digester systems per year. Additional funding has been provided by the World Bank's Community Development Carbon Fund, which purchases emissions savings offset by the replacement of traditional kerosene and wood fuels with biogas.

The subsidies are dependent on the location of the installation, with those living in more remote, mountainous areas receiving a larger subsidy than those in lowland urban areas. They comprise an initial payment to facilitate the construction and verification of the plant. As of 2010, the GPOBA programme had helped to install 10,868 plants throughout Nepal, bringing many opportunities to develop trade activities, providing people with clean energy access and improving sanitary conditions.

Source: de Gouvello and Kumar, 2007

Such initiatives allow the very poor to access high quality stoves which they would otherwise not be able to purchase. An example of an initiative to promote the production and dissemination of energy-efficient stoves in Cambodia, and which has benefited from a carbon financing mechanism and achieved significant impact, is summarized in Box 5.10.

However, like all subsidies, it can be argued that carbon finance distorts markets, and can destroy or limit competitors who do not have access to credits. This initiative in Cambodia, as well as *Ugastove* in Uganda, have been criticized for creating market distortions and causing household cooking technologies to become strongly dependent on carbon markets, jeopardising the sustainability of their use in the long run. As a result some people are starting to call for carbon financing to be used only to help a market get off the ground, or to target the very poor or marginalized, rather than propping up the entire cook stove sector.

Box 5.10 New Lao Stoves: a case of carbon financing for off-grid energy

In 2002, supported by the European Commission, Groupe Energies Renouvelables, Environnement et Solidarités (GERES) and the Ministry of Industry, Mines and Energy collaborated to set up a project promoting large-scale utilization of improved cook stoves, the New Lao Stove (NLS), for urban areas. The main objective of the project was to commercialize NLS in order to promote energy efficiency by reducing charcoal consumption. It was hoped this would then have the impacts of reducing indoor air pollution; mitigating climate change by reducing the emission of greenhouse gases; and creating new job opportunities by producing and distributing the NLS. As of 2010, more than 800,000 units of NLS had been produced and sold, with about 300,000 families enjoying the economic benefits of NLS; time and money saving. Wide utilization of the NLS between 2003 to the end of 2008 has been estimated as resulting in reducing the destruction of more than 400,000 tonnes (600,000 m^3) of wood, equal to 4,500 hectares of forest.

The success of this project resulted in a further grant from the World Bank to establish a carbon facility in Cambodia; the purpose being to access carbon finance for NLS cook stove projects. This new project enabled GERES to conduct studies and develop robust calculations around the baseline and emission reductions achieved by the NLS project. In 2007, the designated operational entity, Det Norske Veritas (DNV), visited the project in Cambodia to approve the emission reductions claimed for the period 2003–2006. DNV verified that the project had avoided emissions of 182,402 tonnes of CO_2. There have been two subsequent visits in 2008 and 2009 verifying reductions of 126,022 tonnes and 192,349 tonnes of CO_2 emissions respectively, under the Voluntary Carbon Standard.

Source: GERES, 2009

Microfinance

Microfinance has recently emerged as one of the most important mechanisms for helping poor households overcome the up-front cost barrier of off-grid energy technologies, and is now used by many organizations in many developing countries. MFIs often have fewer requirements for joining and for providing unsecured loans or loans not backed by physical collateral. MFIs therefore reduce the usually significant barriers the very poor experience when trying to get a loan, as other forms of collateral can be used such as group guarantees (through community cooperatives or SACCOs). Micro credit is now starting to be provided to end users to help them access a range of off-grid energy products, particularly small solar PV and improved cook stoves. The provision of micro credit is usually coordinated by a financial institution and an energy supplier. The loan amounts are often small and the repayments are structured so as to be paid back over a short duration of time, and in line with poor people's income streams (i.e. after harvest time). This type of payment method is also in line with people's already established habits as it is often used for paying for kerosene or charcoal. Microfinance allows peer-group lending mechanisms to develop whereby borrowers form groups to access credit. This reduces risk and increases the likelihood that the loans will be paid back.

Often energy companies cooperate with MFIs to provide loans to the end users, and vice-versa. This close relationship between the financial organization and the energy company is often the key to successful initiatives. An example of an organization which integrates microfinance as part of its business is Grameen Shakti, a Bangladeshi company which produces a range of household off-grid energy technologies, including solar home systems, improved cook stoves, and biogas systems, which can be purchased by end users through a variety of payment schemes, termed 'soft credit', as described in Box 5.11.

Box 5.11 Grameen Shakti's 'soft-credit' business model, Bangladesh

Management: end-user managed
Distribution: commodity market
Scale: national (small to large scale)
Financing: microfinance
Implementing organization: private sector

Grameen Shakti is an enterprise in Bangladesh that manufactures household energy systems and utilizes a microfinance initiative through a 'soft-credit' model which enables poor customers to pay for the technology by instalments. The users initially pay a down payment of between 15% and 25%, which is followed by instalments and a service charge of approximately 8% over the next 24–36 months. The system operates without the use of direct subsidies.

Additionally, the company operates 'micro-utility' models, whereby a user can share the generated load with neighbours and collect a small tariff from them to cover the monthly cost. The after-sales service is covered by the service charge, and includes monthly check-ups while instalments are being paid, and a 'buy-back' warranty system whereby users can resell their systems back to the company in the event of the area becoming grid-connected.

The company's sales have steadily increased over the past 10 years, to the point where over 500,000 solar home systems and 5,000 biogas plants had been installed by 2011. It is estimated that the majority of customers begin to save money within five years of making the initial down payment.

Source: ARE, 2011; Grameen Shakti, n.d.

A slightly different approach is that of Solar Electric Light Company (SELCO)-India, which is a social enterprise established in 1995, offering off-grid solar PV lighting systems for low-income families in southern India (Bellanca, 2012d; SELCO Solar, n.d.). It acts as a facilitator of financing between financial institutions and end users, with 85 per cent of their projects involving consumer microfinancing. They do not get directly involved with the financing operations, including collection and other aspects of the loans, but instead facilitate the relationship by helping the financing institution to understand the needs of the customer, and how the energy systems operate. A very similar model, although using a revolving fund to allocate loans, has been adopted by the Solar Energy foundation (SEf) in Ethiopia and is described in Box 5.12.

Although numerous examples are emerging of successful financial solutions for off-grid energy services, some non-financial companies have encountered problems in trying to implement microfinancing solutions directly on their own. Tough Stuff, a solar PV social enterprise, has experimented with the 'rent-a-light' model in some of the poorest rural communities, but found the

---

Box 5.12 Revolving loan fund financing for solar home systems, Ethiopia

Management: end-user managed
Distribution: commodity market
Scale: national
Financing: revolving fund microfinance
Implementing organization: social merchant bank

The Solar Energy foundation (SEf) was started in 2005, with the aim of forming self-sustaining solar home system franchises following an input of initial capital through revolving-fund microfinance. The foundation saw an opportunity to develop the model in Ethiopia, where at the time, 99% of the population had no access to grid electricity. SEf sells solar home systems for an up-front cost of $80, with monthly payments of $9 for three years thereafter. This is instead of an average monthly spend of $7.50 on traditional fuels such as biomass and kerosene, and results in a much safer and more reliable supply of lighting energy.

As of 2012, four solar service centres have been supporting 2,400 solar home systems, with plans to expand to 50 service centres by 2014, capable of installing more than 50,000 solar home systems per year. It is expected that the initial capital investment in training and service infrastructure will help the local markets grow, with market chain functions being performed by members of the local community at affordable rates. An initial revolving-fund loan of €10 m (US$13 m) was provided by E+Co and Arc Finance, which is being recouped through margins on the monthly billings. It is expected that the loan will become self-sustaining without the need for further subsidization in the near future.

Source: Schützeichel, 2012

---

overheads and collection costs too high to make the model work sustainably, so ended the initiative. Tough Stuff believe that households are often 'risk averse' to taking out loans to pay for solar home systems or other off-grid energy products and services if, as is often the case, their income is unlikely to be significantly boosted by these products, and they do not sufficiently value the social benefits (Van der Vleuten et al., 2007). Poorer households tend to be quite risk averse, as well as being unfamiliar with formal financial services, which can reduce their willingness to access microfinance. However higher income groups will borrow more readily.

Guarantee funds

This commitment by a third party to cover the obligations of other parties in the event of partial or complete non-repayment of a loan, is now starting to be used successfully for off-grid energy services, and might be a useful instrument in the future for achieving Sustainable Energy for All's (SE4All) goal of universal energy access by 2030. A number of organizations have started to work directly with financial institutions to provide loan guarantees. For example, Global Village Energy Partnership (GVEP) International in East Africa are working directly with microfinance companies and localized financial institutions to encourage them to provide loan guarantees to energy entrepreneurs or end users, when they would otherwise be unwilling to do so. This is highlighted in Box 5.13.

Box 5.13 Loan guarantees to stimulate off-grid energy services in rural areas, Uganda

Wekembe SACCO is a cooperative savings and credit society promoting renewable energy in Uganda. GVEP International provided Wekembe SACCO with a partial guarantee to allow them to offer energy loans to 15 small energy enterprises, to help them develop their energy businesses. A female entrepreneur applied for a loan of 2.7 m Ugandan shillings (USh) (US$1,140) under the scheme, and used the money to expand her solar PV-based phone charging business. Her new systems enable her to charge up to 10 mobile phones a day, power four lights for her home, and power a small refrigerator used to chill fruit and sodas, which she sells to a local shop. In addition her children benefit, as the lights enable them to study in the evenings.

With such a loan, a rural-based entrepreneur can earn an extra USh6,000-7,000 ($3) a day and almost halve their monthly kerosene expenditure. Without the loan guarantee, Wekembe would have been unwilling to offer such loans as they would have been perceived as being too risky in a new and untried sector. This approach is now being extended to other SACCOs and MFIs in the region.

Source: GVEP International, 2012

Such loan guarantee arrangements need to be based on good partnerships. The financial institutions need to be sufficiently trained to roll out their energy loan products, and loan officers should be knowledgeable on the energy products when interacting with consumers (the entrepreneurs and end users). In addition, the technical quality of the off-grid energy products needs to be consistent, meeting national or regional quality standards, such as Lighting Africa's approval of high performing solar PV products in East Africa. Effective marketing and distribution strategies also need to be developed to ensure the products can reach the end users, as well as ensuring that the loan repayments can be made. Such loan guarantees can be very important in helping financial service companies to start offering loans for new off-grid energy technologies with a low-risk approach. After a while, once financial services become used to the products, such loan guarantees can be slowly scaled down and eventually withdrawn.

Scaling up loan guarantees for rural areas can be a significant challenge as the financial institutions need to have a good network of branch offices to meet their customers' needs. While localized semi-formal financial institutions, such as SACCOs, provide easily accessible services, their interest rates are often quite high and beyond the reach of the very poor. Developing programmes of loan guarantees takes a lot of time and effort, and some argue that they are not a very effective use of scarce financial resources. However, if successful, loan guarantees can be an important tool in assisting off-grid market chain actors to expand their services and products to the rural poor through a lower reliance on grants and subsidies, and a gradual move towards a purely market-based approach.

Informal financial services

In addition to structured financing, informal methods of sourcing credit for off-grid energy services are very common in many developing countries. These informal methods involve borrowing from relatives, friends, and moneylenders, and entail few formal requirements or need for proper accounting. However there are significant disadvantages, such as limited fund availability, or high interest rates (especially from moneylenders). One of the common informal institutions are the rotating savings and credit associations (ROSCAs) whereby members contribute a set amount of money regularly and then the total, or part of the total raised, is loaned to one of the members at each meeting until all the members have been covered. Such schemes generally have few requirements and often the interest rates or fees are nominal. Due to the informal nature of these community-based models there is a risk that participants could lose money if the other members fail to make their contributions; however, due to the high levels of trust and peer pressure present in such communities, they tend to work well.

In addition, remittances received from relatives working in developed countries, or between rural and urban parts of developing countries, are starting to be recognized as able to play a significant role in off-grid energy access in developing countries in the future. Some energy companies are starting to experiment with marketing targeted at wealthy individuals, encouraging them to purchase certain off-grid energy technologies for their poorer relatives, such as improved cook stoves or solar home systems.

Capacity building

Capacity building needs to take place at different levels, from the end users (including those who do not have access to certain energy services), to the organizations and institutions involved in the design and implementation of all parts of the off-grid market chains, as well as those involved in local and national energy policies. The latter includes both public- and private-sector organizations, as well as the financing institutions that can help to support off-grid energy services.

It is important for capacity building to take place throughout the market chain, as highlighted by a programme from Bangladesh which focused on the promotion of off-grid solar PV technologies. It was only after capacity building was provided using a holistic approach – for marketing, promotion, and quality control of the products, as well as for the provision of soft loans for the end users – that the programme was able to catalyse the market demand for off-grid solar lighting products in the country and reach real scale (IFC, 2012).

One of the major challenges for off-grid energy delivery has been their long-term financial viability. Capacity building is able to play an invaluable role in improving this. Often programmes have failed because of the lack of infrastructure, but more often it has been due to the absence of skilled people to deliver the services and products, and then maintain them, as well as the lack of appropriate channels of financing that fully understand the markets. As highlighted in the previous section, many financial institutions do not have specific energy portfolios, and there is a real need for capacity building for these institutions. This will help enable them to develop energy loan products for a range of off-grid energy technologies (solar PV, improved cook stoves, biogas, etc.) together with the energy enterprises. Donor-funded programmes and NGOs are often the major sources of capacity-building initiatives. Their support may include building up knowledge and skills, not only for the local institutions directly involved in the off-grid energy technologies, but also for those who might help deliver such technologies (e.g. local academic institutions, research institutions, private-sector companies, as well as public institutions and communities). The work carried out by the EU Energy Initiative-Partnership Dialogue Facility in Ethiopia (detailed in Box 5.14) highlights this point.

Box 5.14 Capacity building for lending institutions in Ethiopia

An initiative to build capacity in Ethiopia for the delivery of off-grid electrification was implemented through collaboration between the Ethiopian Ministry of Mines and Energy and the EU Energy Initiative-Partnership Dialogue Facility (EUEI-PDF).

The project worked to strengthen ties between the Rural Electrification Executive Secretariat and different governmental levels in the country through the development of a rural electrification master plan which was successfully disseminated among all government departments, leading to greater cooperation. In particular, the requirements of the national plan were communicated to regional energy bureaus and finance bureaus, in order to ensure that the regional requirements could be realized within the overall framework.

One particular communication tool that has been adopted on a national scale is the geographical information system database and GEOSIM simulation tool that allows standard planning procedures to be decentralized and implemented autonomously by regional departments. This includes the provision of off-grid energy technologies in rural areas of Ethiopia.

Source: EUEI-PDF, 2011

The lack of capacity at the institutional level, especially in the public sector, is a major constraint and still needs to be addressed. In particular, clear and targeted approaches are required to develop policies and programmes for scaling up the off-grid energy sector. Many rural energy or rural electrification agencies are still reliant on donor support to help build their capacity (including in technical support), and this is likely to continue for some time in the future. Policies that favour off-grid energy technologies need to be designed in cooperation with other sector institutions such as the ministries of health, environment, water, and land use. This cross-ministry coordination is a vital prerequisite for overcoming energy poverty. The use of national-level statistical data and analysis for a range of energy resources is needed to ensure that off-grid energy delivery systems can be effectively scaled up, especially to poor or rural remote areas. The lack of confidence in understanding this market among public-sector legislators and decision makers is often a significant hurdle to the design and implementation of appropriate policies, and is something that needs to be addressed in the future. In addition, public institutions need to have sufficient skills to roll out any large off-grid energy programmes, including the monitoring of their impacts to ensure their benefits can be captured and lessons learned and shared.

Many capacity-building initiatives have been developed as stand-alone activities, and a more interactive platform involving a range of market chain actors is often non-existent, especially for off-grid initiatives. Many poor or rural populations do not have effective communication channels, such as the use of social media or internet access, to gain the information they need about the range of off-grid energy services, and most energy companies developing these services still highlight this as being one of the main barriers that needs to be overcome. There is a tendency for the more interactive knowledge-sharing platforms to only be directly accessed by more affluent end users in urban areas, rather than by those that really need them. Thus, awareness-raising has become a common capacity-building tool used by many to try to reach the rural users of off-grid energy systems. This is expanded on further in the next section.

Awareness-raising and marketing

Awareness-raising and marketing have traditionally been used by many non-profit organizations to try to help increase energy access for those living in peri-urban and rural areas of developing countries, as well as for the urban poor. These activities are often carried out through informal education activities and are usually short lived, specifically focused on project interventions or specific product information. They are not carried out over an extensive time period and for a range of energy services. Awareness-raising has been shown to be an effective way of creating demand for a product or promoting participation among users. It can be especially impactful if it can be rolled out to a large community. Some initiatives, however, have assumed awareness-raising for the poor is a rather simple task and have therefore not been well prepared, leading to mediocre results.

As mentioned in the financing section, market intelligence directly related to the energy needs of the rural poor, particularly through off-grid energy technologies, is still not well-researched. For any customer, even if they are poor, the accuracy of the information they can access is essential in allowing them to make informed decisions on whether to buy a product or not. In East Africa, the EU Energy Facility funded the Developing Energy Enterprise Programme (DEEP), which has developed a number of key lessons in relation to awareness and marketing of off-grid energy services. Many energy entrepreneurs were found to be lacking the appropriate marketing skills to effectively promote their products, and struggled to recognize the different market segments that existed in order to customize their products. This combined with the end users' lack of awareness of available off-grid energy technologies, particularly in rural areas, heightened the need for entrepreneurs to engage in effective marketing activities. DEEP facilitators focused on the main marketing principles of product, price, promotion, and distribution. They organized group networking and information-sharing sessions to allow energy entrepreneurs, customers, suppliers, and other stakeholders from different regions to come together. These sessions were very successful, with energy entrepreneurs reportedly gaining customers and buyers in new areas. Such participatory information-sharing helps actors right across the off-grid energy technologies market chain, leading to win–win for all organizations, as well as a much better service for the end users (GVEP International, n.d.).

Larger energy enterprises such as D.light and Tough Stuff have been working to encourage local shop owners to sell on their products in their shops and other public spaces. They have been helping them raise awareness of their products in an inexpensive way, in order to generate increased general awareness of that technology, in this case solar generated electricity, as well as their own particular brands. It is hoped that as awareness of particular off-grid energy technologies increases and as company images are established, this will become progressively easier. Other factors that influence end users' willingness to adopt a new technology include their general exposure to the products through marketing strategies, such as public demonstrations in markets, as well as the public reception to the products. News generally travels quite fast in rural areas of developing countries. After buying a poor quality cheap solar product once or twice, and seeing it break, people can quickly lose confidence in any solar products.

In Nigeria and South Africa people readily pay for expensive diesel-to-power generators but expect state-run energy services such as grid electricity or mini-hybrid grids to be cheap or free. Cultural attitudes to the potential of solar energy are far from positive in these countries, particularly after many failed public solar projects (e.g. the Niger Delta Development Commission's failed solar water pumps, or the solar street lights in Port Harcourt). Solar products have a high perceived value in areas where there is no or intermittent electricity. Products that are manufactured to a high quality are robust, generally require very little maintenance, and generally encounter good levels of acceptance; however, when users have a bad experience from a poorer quality product this can result in negative perceptions of the whole sector. This is why quality control standards and awareness-raising around the differences between high and low quality products is so important. Quality control issues can also be overcome by the issuing of warranties, and selling products through people who have established trust with the local communities (e.g. informal sellers, teachers, and trained village entrepreneurs). D.light often sell their products through schools, and early results have been positive, with in some cases, 70–80 per cent of students in a class choosing to buy them. Solar Sisters has been developing local markets for their off-grid energy technologies through the use of the Avon Cosmetics social sales model, and the establishment of local saleswomen and respected advocates (Aylett, 2010).

Enabling environment

The enabling environment is of critical importance in allowing off-grid businesses, in particular the development of distribution networks to rural areas, to flourish. As many off-grid energy technologies (particularly solar PV, but also increasingly a range of high performance ICS and clean, packaged fuels such as kerosene and LPG) are currently imported into developing countries, the national-level importation and tax regimes are of particular relevance, as well as the standards and quality control, as mentioned in the previous section.

Policy and regulations

Overall, each country needs to develop their own appropriate policies based on their localized or regional contexts in respect to off-grid energy delivery. Middle-income countries, where off-grid energy technology markets have a lower potential, may need different policies as compared to low-income countries that have a much higher potential for such markets. This is due in part to the much lower rates of grid electrification in low-income countries. In many Latin American and Caribbean countries, off-grid lighting markets are remotely located with very low densities making it very difficult for them to be served by purely private-sector companies. Mitchell et al., 2011, categorizes four types of policies based on fiscal incentives, public finance, regulation, and access policies. One of the important access policies is to allow the two-way flow of electricity between grid distribution companies and those households who generate electricity on their own. Although this concept is gaining strength in developing countries, it still needs a fair bit of regulation in poorer rural regions where lack of infrastructure is often an additional hindrance.

With fossil-fuel prices on the rise, the balance between fuel demand and supply is going to be pivotal in determining prices and access. Geopolitical factors are likely to influence the way fuels will be used and distributed, and policy has a strong role to play at this level. Some packaged fuels, such as LPG or kerosene, are likely to be subject to government regulation, and subsidies may be applied at various points. These could be at the initial purchase of the raw fossil fuels, through to the processing and distribution processes, and on their final selling price. For example in Ghana, kerosene for household use is subsidized at 23 per cent below the price in the mining sector, and an additional primary distribution margin, a levy for the Unified Petroleum Price Fund (UPPF) and marketeer charges, are added to encourage transportation and regular supply to remote locations. This encourages uniformity of prices and availability. In addition, all petroleum products are subject to a levy by the Energy Fund to support R&D and to promote Ghana's natural energy resources, particularly renewable energy (Lighting Africa, 2012). However, another report on Asia, by Lighting Africa, showed that kerosene subsidies could be a major challenge to the development of the solar off-grid lighting market. For example in India and Indonesia, significant subsidies are provided by governments for conventional fuels, including kerosene for domestic usage, thereby making it a big challenge for solar companies to convince consumers to switch to cleaner lighting products (IFC, 2012). Some policy interventions that were initially intended to serve the poor have been vulnerable to 'leakage' towards other markets such as transport or export. An example of this is the household ethanol market in Ethiopia (see Box 5.15), where locally produced ethanol (processed within state-owned refineries) was initially introduced as a clean household cooking fuel, but was later diverted by a sudden change in policy, to the transport sector. It has taken almost one year for the state to re-evaluate its priorities; and has currently promised to ensure that a significant supply of ethanol will be devoted to the cooking sector to meet the future needs of millions of families in the country (Practical Action Consulting, 2009b).

Box 5.15 Gaia ethanol programme, Ethiopia

Management: NGO/private/public
Distribution: decentralized
Scale: local (small scale)
Financing: donor
Implementing organization: International cooperation

Project Gaia is active in promoting ethanol as a household fuel and has been developing alcohol fuels for household energy for some years. The model addresses the whole cooking value chain starting with the production of ethanol fuel, followed by the stoves to use it with, and then their marketing and distribution. Ethanol is most commonly derived from sugar cane, cassava or other biomass sources. The first projects were implemented in Ethiopia in refugee camps as well as in the capital Addis Ababa. The fuel was provided by state actors and underwent an interruption in supply during 2011 when ethanol fuel was diverted to the transport industry. As a consequence of re-evaluation of priorities at the government level, regular fuel supply was resumed in 2012.

A project has followed in Haiti and another spin-off initiative is underway in Mozambique (see Box 5.3). In 2007, the Gaia Association began working with a private-sector partner to facilitate local manufacture of the stoves (CleanCook), to reduce the cost of the stove to Ethiopian consumers. The stoves are of Scandinavian design and were previously produced in Europe. Micro-distilleries to produce the ethanol are being built or planned in Ethiopia, Mozambique, and Haiti.

The potential for the diffusion of ethanol as a cooking fuel is believed to be huge, and Project Gaia is hoping to be able to replicate the way in which the kerosene market grew around 100 years ago, but faster. As demonstrated by the Ethiopian case, the key threats to diffusing ethanol for household uses are the export and transport markets which are looking at blending ethanol as fuel for car engines, influenced by Europe's biofuel appetite as well as the African transport industry, and inspired by Brazil's ProAlcool model.

As in the case of LPG, ethanol represents a superior alternative to biomass in terms of its cooking convenience, its impact on health, and its acceptance by households. It also has a positive outcome on deforestation. Moreover, as the ethanol value chain is a local process at each step, it provides clear economic advantages for the local population employed in agriculture, industry, and distribution. Finally, the possibility to locally source the fuel means that remote communities can farm, process in micro-distilleries, and directly use the energy source without depending on national or international programmes.

Source: Bellanca, 2012b

Many factors need to be evaluated in designing effective policies for off-grid energy systems. In the case of ethanol in Ethiopia, while household energy markets can provide a nice return for ethanol manufacturers it might not be as lucrative as the export market or for the blending of auto fuel. On the other hand, Ethiopia has a serious deforestation problem, as well as an unsustainable and unhealthy biomass-based household cooking sector, and so needs to try to meet the demands from its citizens for clean cooking fuels. Moreover, the export market presents risk factors for Ethiopia as it is landlocked and the cost of exporting is high. It is relatively small player in the global export market for ethanol and has very little leverage in commodity pricing. All these elements, and others, need to be thoroughly analysed when developing and instigating policies for encouraging one type of energy delivery model over another.

Many developing countries, particularly in sub-Saharan Africa, have tried to control the over-reliance on biomass for household cooking, which has led to numerous negative impacts such as the depletion of forests, by making the production and trade in charcoal illegal. However, as there are often no realistic alternatives to charcoal in these countries, particularly in urban areas, this form of absolute regulation has not been effective, due to the practicalities of costs, law enforcement, and resources required. In reality, the trade has continued but has shifted to the black market, often resulting in corruption, conflict, and exploitation, particularly in areas such as eastern Democratic Republic of Congo (DRC) where charcoal trading has been linked with the funding of guerrilla warfare. In some countries, such as Kenya, this has started to change, and the production and trade in sustainably produced charcoal is no longer illegal. Although this change has not yet been effectively communicated or regulated, Practical Action has started working closely with Kenya Forestry Service (KFS) to produce charcoal policy handbooks to communicate the new regulations to all the market chain actors (producers, distributers, and charcoal sellers), to try to improve the regulation of the market (Practical Action Consulting, 2012).

The high level of investment required to achieve the IEA's aim of Universal Energy Access for All by 2030 cannot be met purely by government and donor funding. As a considerable part of this investment is expected to go to off-grid systems, the creation of regulatory frameworks that encourage private-sector investment – particularly in decentralized, off-grid and household delivery models, which have traditionally been less 'attractive' to private investors (OECD/IEA, 2011) – is an important aspect of policy intervention. Box 5.16 highlights an example of this from Mali, where activities have been underway to increase investment in off-grid energy technologies. Several policy plans have been introduced in developing countries to attract investment in a range of energy technologies.

One of the most effective instruments for governments in steering investment towards off-grid energy systems is through taxation. For example, many countries have waived import duties on solar components, electrical

---

Box 5.16 Government policies to encourage investment in off-grid technologies in Mali

The development of national policies to stimulate investment in off-grid energy has taken place in many countries. This example looks at Mali, where in 2008, the Malian Government adopted the 'Renewable Energy Development Strategy', which was followed up by the 'National Biofuels Development Strategy'. Part of these strategies involved the creation of a National Agency for Investment Promotion, which had the specific aim of attracting foreign investment into rural energy delivery programmes. Prior to this in 2003, the Agency for Household Energy and Rural Electrification (AMADER) had been set up to provide capital subsidies of up to 80% for infrastructure investments. AMADER also provided limited amounts of technical assistance and capacity support to private companies. Approximately 70% of Mali's population is rural, with large potential for development of bioenergy. The policies introduced by the government have addressed many landholder disputes and inconsistencies in land access, creating an environment much more conducive to investment.

Source: Aron et al., 2009

---

components or raw materials for manufacturing, which has provided a boost to these markets. In Ghana, only solar panels are exempt from duties but still face a 12.5 per cent VAT and fees of 3.5 per cent, while solar system accessories have no tax exemption. Off-grid solar lighting products receive no tax exemption, and this is an area where governments can initiate change, increasing access for poor, and/or rural consumers by ensuring the products are exempt from all taxes (Lighting Africa, 2012). Interestingly, a study for IFC concluded that policies and subsidies that strengthen the whole value chain are more effective than a direct subsidy intervention that reduces just the end consumer price (IFC, 2012).

Socio-cultural context

Appropriate design

Experience shows that stoves designed in laboratories without extensive fieldwork experimentation are less readily adopted than those that are specifically designed with the local users, to meet their cooking preferences (Barnes et al., 1994; Cecelski, 2004). In some cases there has been strong, cultural rejection of particular cook stoves that were not appropriately designed (Vermeulen, 2001), with a study from Uganda showing the importance of issues relating to local cooking taste being critical to stove uptake (Wallmo and Jacobson, 1998).

It is important to consider the cooking fuel along with the stoves – people may have a strong preference for continuing to use traditional wood fuel due to its availability and low, or zero, cost compared to other fuels (such as LPG or kerosene, which are generally much easier to cook on), and will therefore be unwilling to adopt alternative options based on other fuels. Solar cookers are essentially a western invention designed to try to replace fuels that need to be collected or purchased, with those provided by the sun's rays and which are free. However they do little to take into account the needs of their target populations as they force the user to cook in the sun, and only allow the first meal of the day to be cooked by midday and the last meal in the afternoon, rather than in the early morning and after sunset respectively, when they are usually required. They have been found to be useful in very specific situations such as for lunches in schools and in prisons, but not typically for households.

Successful programmes have managed to respond to local needs and preferences, sometimes even at the cost of making compromises on donor objectives. In Sri Lanka, the stove finally chosen was not the most efficient stove on offer, but it was more efficient than previous practices. The design still resembled the traditional model, and the stoves were produced using local resources and skills. This outcome supports conclusions from elsewhere indicating that designing a more efficient stove that strongly resembles a more traditional one can work well (Barnes et al. 1994; Karakezi and Walubengo, 1989).

There is an important gender aspect to be considered in the preferences of households. The extent to which women's priorities are valued and their needs understood is likely to affect the rate of adoption, particularly when women do not have the power, or access to finance, to make new purchases. For example, Clancy and Skutch (2004) suggest that stoves may have had more success in China because women are directly engaged in the cash economy, therefore have clear incentives to save time. Preferences should also be considered in the context of household power relations, for example whether women have control over their labour and can influence decision making (Cecelski, 2004).

Open competitions have been shown to be effective in developing designs of off-grid energy technologies suited to local needs, both in Sri Lanka and China. It is worth noting that China's National Improved Stoves Programme was a top-down programme, but there was still encouragement for local variation and innovation to find stoves well suited to people in different areas (Wilson et al., 2012). Solar product companies, such as Barefoot Power and D.light, consider all aspects of the business model (the product, affordability, the distribution models, marketing, and the warranty), and aim to ensure these are in place in each market they enter; however there is still a need to have a strong presence 'on the ground' to be able to understand why a product is, or is not, successful (Wilson et al., 2012). Off-grid energy products have been found to be well liked by customers, but this does not necessarily lead to significant sales if issues, such as the distribution model or warranty guarantee, have not been overcome. As mentioned previously, the willingness of households to adopt off-grid energy technologies is also affected by the likely probability of the arrival of on-grid electrification, particularly for solar PV systems. People living in areas close to the grid, or within estimated expansion plans, may be unwilling to pay for relatively large investments such as solar home systems. In these cases, off-grid electricity initiatives need to be aware of any government plans to expand the grid, and, in turn, governments need to be open about their realistic on-grid expansion plans.

End-user preferences

For a business or development intervention to be sustainable and implemented over a significant timescale, it is important to ensure that the goods and services meet the end users' preferences and expectations, particularly regarding the affordability, functionality, durability, appearance, and status value of the conversion equipment and appliance technologies. This is of particular relevance for cook stoves and small electricity-producing technologies (e.g. solar PV units), where the participation of the end users in their design is vital to their long-term usage and ongoing sales. This is highlighted by the *Anagi* improved cook stove example from Sri Lanka (see Box 5.8) which was designed to meet the needs of local cooking practices, and has reached sales volumes of several million through local production centres, albeit after an extensive period of donor support.

Another well-known off-grid energy delivery model is the dissemination of solar home systems in Kenya. This has proven to be a sustainable and viable market-led solution to delivering electricity to households not connected to the grid. However, as they were not designed to specifically meet the needs of the very poor, they have not been taken up at scale or provided the range of 'developmental' benefits hoped for (Jacobson, 2007). In contrast, many other small solar PV products have benefited the poor, such as those developed by Tough Stuff and Greenlight Planet, as they have been specifically designed with the very poor in mind. Their simple but appealing designs, ease of usage, low maintenance requirements, and durability, in addition to being much more affordable than solar home systems (due to significantly lower capacity), have made them very appealing to their target market. These off-grid products are not able to offer the full benefits of grid-electrification, but they are able to make a crucial difference to households with no access to modern energy, and importantly, do not draw significant public resources away from other developmental priorities.

Such evidence shows that when designing off-grid energy technologies, the price of the units is not the only consideration. Durability and aesthetics are also important factors, as well as the addition of premium functions such as mobile phone charging, which has really helped the uptake of some models of solar PV lanterns. Users' attitudes to looks and functionality vary culturally and one of the challenges is to ensure that the products appropriately fit the local contexts while still being affordable to produce and sell. The development of close relationships with the potential customers is another very important part of any R&D phase in order to determine their priorities, tastes, and needs, as well as their price sensitivity.

Companies such as Barefoot Power, Tough Stuff and D.light have emerged as dynamic and leading social enterprises. They supply a range of locally appropriate and affordable solar PV products to local markets in many countries in Africa, due to their close involvement with the end users. They are enabling low-income groups (with typical daily earnings of $1–5) to obtain small electrical lighting systems. This allows them to move away from a reliance on candle or kerosene lighting or expensive generators, to safer, healthier, and more sustainable alternatives, that provide better quality lighting as well as significant savings in the long term. Users are often able to gradually upgrade their off-grid systems, by starting with a very small solar panel and LEDs, and then gradually increasing its capacity over time. Some companies, such as Solar Now in Uganda, allow users to trade their old solar PV systems in when they're able to afford to upgrade to a larger system. These companies focus on ensuring high quality and durability to deliver a positive energy experience for the end users. They also encourage initial intensive engagement with their target customers during the product design, and through marketing and awareness-raising activities. A number of improved cook stove companies are trying to replicate this approach.

CHAPTER 6
Conclusions

This book has provided an analysis of some of the issues surrounding energy poverty and the importance of energy access as a basic human need. Using Practical Action's participatory market systems development framework, the book looks at each delivery model's market chain, the supporting services, and the enabling environment, which together cover the infrastructure, policies, trends and regulations surrounding the energy system, and the socio-cultural context of the end users of the energy services. The core of the book is concerned with how this framework can be used to analyse, understand, and design energy delivery models at the scale of on-grid, mini-grid, and off-grid. Energy practitioners can use this information, particularly lessons learned, in the development and design of future models in order to provide universal energy access to a range of services and reduce energy poverty in developing countries.

Keywords: energy delivery model, energy poverty, participatory market systems development, scale of delivery and sustainable implementation

This book has attempted to provide an analytical description of some of the issues surrounding energy poverty and the importance of energy access as a basic human need. The book has looked at the way in which projects delivering energy services have evolved historically and considered the roles and partnerships of the actors involved. It has presented a more systemic approach to understanding and classifying energy delivery models to try to overcome the problems associated with their sustainable implementation, management, and maintenance. Energy delivery models have been divided into three broad segments: the energy market chain, the supporting services, and the enabling environment. These segments are based on Practical Action's participatory market systems development (PMSD) framework. This approach has then been categorized into three scales of delivery, namely on-grid, mini-grid, and off-grid energy systems. The core of the book is concerned with how this framework approach can be used to analyse, understand, and design energy delivery models at these three scales. The book provides information and guidance to energy practitioners in the development and design of future models in order to provide universal energy access to a range of energy services and reduce energy poverty in developing countries.

For an energy delivery model to be successful, energy practitioners need to develop a detailed understanding of each segment of their energy delivery model, including all the actors and organizations involved, and how they

http://dx.doi.org/10.3362/9781780447612.006

interact and affect each other. In addition, the roles of the three scales of energy delivery models in supplying a range of energy services need to be understood, and how they can complement each other to effectively deliver a country or regional energy strategy. It is not possible for all energy services (including lighting, heating, and productive services for households, community services, and businesses) to be delivered through only one scale of energy delivery model. While the on-grid model focuses largely on electricity provision for urban and peri-urban areas, mini-grids are common for peri-urban and rural areas, and off-grid delivery caters mostly to rural populations. However, solar home systems and improved cook stoves are also becoming increasingly important for urban and peri-urban consumers.

This framework approach has been developed through an analysis of numerous case studies from published literature as well as from the authors' own experiences. This diverse range of examples illustrates energy initiatives that have been implemented on the ground by governments, the private sector, NGOs, and development banks, often in coordination. The examples have been used to explore different aspects and best practice lessons around the design of energy delivery models, ranging from the technologies involved, the management and maintenance models, the financial services, and the socio-cultural environment. In addition, the mapping of each energy system, ideally through a participatory approach involving stakeholders from all the key segments, allows a graphical representation of each energy market to be developed. Such market mapping techniques can be applied to all scales of energy delivery models to enable a better understanding of the three main segments, and the interactions between each actor. This hugely improves the ability of energy practitioners to identify the blockages and gaps in the delivery of energy services to the end users and to design initiatives to overcome them. By initiating such cooperation between all involved actors, this method promotes empowerment, interaction, and communication as well as the promotion of a systemic approach to delivering a range of energy services. This guided process of analysing each stage of an energy market chain (from the initial energy sources, through to their processing and energy conversion equipment) as well as the required appliances (including the identification of the roles of each market chain actor, the surrounding environment and the demands of the end users) enables energy practitioners to design and implement effective energy delivery models.

A range of challenges have been identified that might impede the effective functioning of an energy delivery system. These include a lack of capacity in the identification of appropriate energy resources, or in the production and design of the conversion equipment or end user appliances. They also include the inability of the end users to afford the services, or to value the services highly enough to prioritize their purchase. Appropriate financial services that allow the poor to access certain energy technologies are also very important. In other examples, challenges have been identified within the enabling environment, such as national governments not being able to create suitable climates for a range of energy services, particularly in terms of reaching those in very remote locations or those who are very poor and marginalized. Supportive policies and regulations might include tax incentives, support for innovation, or quality standards for energy technologies.

Success of energy delivery models

The examination of relevant case studies on the three scales of energy delivery models – on-grid, mini-grid and off-grid – has shown that each scale of delivery models is important, and complementary for achieving universal access to a range of energy services, provided that the three segments of the energy systems are effectively functioning and complement each other. A particular energy market chain is likely to be more successful if the enabling environment is effective and the necessary supporting services are in place and well designed. Similarly, the enabling environment and energy market chain can be strengthened through the right supporting services. For example, policy makers need adequate knowledge and skills to design appropriate policies and implementation frameworks, which can be provided through research and analysis. The end users of the energy systems (communities, businesses, and households), need to be made aware of their particular features, and trained to better manage and maintain the systems, or to more effectively use the services they receive.

Creating the right enabling environment for an energy delivery model might not alone lead to significant impact, particularly if there is limited local capacity to deliver and utilize the services provided along the market chain. For example, several feed-in tariff policies have been developed in a number of developing countries, but have shown very little sign of effective uptake. This has been for a range of reasons such as irregular policy implementation with respect to tariff setting or levels of subsidies; lack of adequate financing; lack of promotion and support for the private-sector actors (and state utilities where they are actors in the energy market chain); and compensatory issues with land owners. The key lesson is that a single policy may not create the right enabling environment for the uptake of an energy delivery model and may need a number of other enablers to make it successful.

Through the analysis of the case studies and experiences presented in the earlier chapters, a number of key criteria have been identified for maximising the success of each scale of energy delivery model, presented in Table 6.1. It is important that each of these is analysed in detail, and if they can all be addressed, the energy delivery model will have the highest probability for replication and success. The exclusion of one or more of these criteria, may not lead to the failure of the energy delivery model, but will limit the likelihood of its long-term sustainability.

Table 6.1 Criteria for success of each scale of energy delivery model

Delivery model scale	Market chain	Enabling environment		Supporting services
		Socio-cultural context	Policies, regulations, and governance structures	
On-grid	• Availability of adequate resources (e.g. processed diesel, hydro or wind resources, geothermal, etc.) • Producers/developers/distributors in place (for electricity as well as appliances for consumers) • Existence of grid and potential to extend it where feasible • High willingness of utilities to find innovative ways of reaching the poor where it is logistically possible • Regular payment systems initiated	• Willingness and ability to pay by poorer groups • Aspiration to connect to the grid exists • High level of consultation with poor end users, including local ownership if possible • Clear process for ensuring access for poor segments • Acceptance of end users to pay regularly for on-grid energy services	• Utilities encouraged to work with financial institutions, including micro-credit • Regulations for selling electricity via third parties (entrepreneurs/cooperatives/franchisees, etc.) • Global trends and policies regarding fossil-fuel usage (applicable to mini-grid and off-grid as well)	• Awareness and marketing of on-grid energy services to better highlight their development benefits and how the poor can better access them affordably • Financial institution providers willing and adequately trained to engage in on-grid energy delivery, including having services available in suitable locations. Simplistic, uncomplicated design for payback – effective use of mobile payments, etc. • Grants, financial incentives, and technical assistance provided to utilities to allow them to provide specific services to poor end users
Mini-grid	• Adequacy of resources. • Strong technical and financial capacity of developers (including product quality) • Provision of ongoing maintenance services • Regular payment systems and appropriate tariffs to cover operation and maintenance essential • Management – public, private, community, public–private	• High level of skills and capacities to maintain systems • High level of ownership and management • High community cohesion • Users' willingness and ability to pay for services • User acceptance of technologies	• Grid coverage is absent, unlikely to reach in the near future, or system predisposed to feed in to grid • Rural electrification policies and strategies in place and ongoing political, technical, and financial support from a range of government departments • Presence of good infrastructure, including communication and roads	• High level of skills development for feasibility, instalment, management, and maintenance • Financing systems and incentive design required for up-front capital costs

Table 6.1 Criteria for success of each scale of energy delivery model *(continued)*

Delivery model scale	Market chain	Enabling environment		Supporting services
		Socio-cultural context	*Policies, regulations, and governance structures*	
Off-grid	• Adequacy of energy resources • Product developers and distributors in place with good marketing and management structures • Availability of adequate maintenance services • Products well-researched and developed to meet quality and appropriate designs	• High level of skills and capacities to maintain systems • Users' willingness and ability to pay for services • Local user awareness of benefits of energy services • Local user acceptance of technologies	• Well-developed infrastructure to ensure technologies and services can be distributed to all areas of the country • Tax reliefs on product imports or production • Standardization and quality control in place • Financial incentives available, particularly during the initial R&D and piloting development stages	• Availability of capacity-building institutes or personnel in country • Presence of financing mechanisms for end users, product developers, distributors, etc. • Financing institutions have adequate branch offices and personnel adequately trained • Institutions support R&D for off-grid products.

Key messages

Drawing from the examples of energy delivery models that have been implemented globally by various actors, the authors have formulated some key messages from the three scales of energy delivery, which can serve to guide energy practitioners. These are summarized in the following section. This book has been written from a development perspective, and so with a focus on the central requirements of the design and implementation of energy delivery models being to increase energy access to achieve a range of wider social and environmental benefits. As outlined in Chapter 1, the supply of energy services is seen as instrumental to the delivery of a range of social benefits, including improved health and education, increased income through productive uses of energy services, and a move towards low-carbon energy resources through promotion of clean technology.

Energy market chain

Central to any energy delivery model is the market chain, characterized by a range of actors involved in the activities required to capture and process the energy resources, as well as develop and install the technologies, distribution networks, and management models. Successful experience indicates that robust market chains are generally quite extensive, integrating the energy producers at the start of the chain, all the way through to the end users at the far end of the chain, with a number of organizations and entrepreneurs, working alongside through interactive and positive relationships. The market chain actors need to understand the range of supporting services that are available as well as the enabling frameworks that are in place for the delivery systems to be effective and sustainable. While national on-grid systems have traditionally been sought after by end users for their high quality of service and relatively low costs, it is becoming increasingly clear that on-grid systems alone will not be able to provide the range of energy services required to eliminate energy poverty. In addition, the era of cheap oil is coming to an end and the prices of on-grid energy delivery are likely to gradually increase over time. This is notwithstanding the gradual global trend towards low-carbon development which is already impacting on developing countries. In addition, the on-grid infrastructure required to reach households in more remote and sparsely populated locations, will be prohibitively expensive to build and maintain and therefore not a realistic option. This will ensure that mini-grid and off-grid energy delivery systems will be an integral part of the future energy plans of most developing countries.

Case studies from this book have highlighted the need for utilities, both public and private, that are willing to design and develop on-grid delivery systems to meet the needs of the poor. They also need to actively engage with a range of innovative financial models with a number of financial institutions, including grants, venture capital, revolving funds, appropriate payment methods, and microfinance. The success rates of on-grid energy systems that meet the needs of the poor are generally still very low, but it is important that such systems are prioritized in the future, as urbanization continues to increase and with it the number of poor customers being served by on-grid energy systems. The electricity sector could learn lessons from a number of off-grid energy companies who have embraced the delivery to the base of the pyramid, developing specific product portfolios just for their target audience.

It is likely that mini-grids will form a more important part of the delivery of energy services in the future. They can be set up in very remote areas using a range of local energy resources (more than one energy resource is used within hybrid systems), and are often more cost-effective than on-grid extension. In addition, they are still able to deliver energy to a number of households and so can take advantage of economies of scale and specialized management and maintenance teams, which might not be possible for off-grid energy systems. Finally, mini-grids can be more specifically designed to meet the local users' needs, than much larger and more generic on-grid energy systems.

Some energy delivery models take a more integrated approach by aiming to eradicate poverty rather than working around it. These models are designed to proactively mainstream activities that use energy services in a productive way to increase incomes, rather than assuming such activities will flourish on their own once the energy becomes available. Off-grid energy market chains aimed at servicing the poor often operate in uncertain or unproven territories, either because the services and products are new to the markets, or due to poor infrastructure, remote locations, and low customer purchasing power. Off-grid energy market chains have been shown to be most successful when existing networks are utilized and innovative approaches to ensuring the financial sustainability of the systems are prioritized.

Supporting services

Financing

A common thread running through all scales of energy delivery models is the importance of financial services. These are essential for the initial piloting and implementation of energy systems through to grants and loans, and also affect the affordability of the energy conversion equipment and appliances. A number of innovative strategies have been adopted to overcome the up-front cost barriers for low-income populations, as well as the maintenance costs of generation, distribution, and technology infrastructure. These challenges have been met through designing innovative financing models such as mobile phone payments (e.g. M-Kopa) and scratch card energy top-ups. Stepped payment levels for on-grid energy usage allow the poor to cheaply access low amounts of energy. Community savings schemes and tailored subsidy schemes are also very important.

When implementers of energy delivery systems source financing, they need to be aware of the range of types of financing organizations. Social impact investors in energy delivery models are usually prepared to receive lower financial returns over longer pay back periods, but expect high socio-environmental standards. In contrast, investors that are focused on supporting businesses that provide energy services to low-income consumers need to appreciate that their return on investment will most likely be lower than other private capital investments. Failing to understand these realities will lead to false expectations and eventually to failure, evidenced by the recent restructuring of E+Co.

Financing and investment can come from a number of sources but it is clear that a wide range will be needed to scale up the delivery of on-grid, mini-grid, and off-grid energy systems to meet the needs of the poor, and achieve significant impact in the elimination of energy poverty. Businesses focusing on delivering off-grid energy technologies, such as solar lanterns or improved cook stoves, are often well suited to the needs of the poor. However they are often not included in the energy portfolios of financial institutions due to their high transaction costs and often informal market chains. These challenges have sometimes been overcome by simplifying and optimizing the financial transaction services through the use of information communication technologies (ICT), or by creating specific packages of credit designed for off-grid energy technologies.

A long history of government and donor financing of energy delivery models, such as for the *Anagi* stove in Sri Lanka, has shown that long-term commitments to developing sustainable delivery models are requirement to achieve positive results in reducing energy poverty. Programmes that provide financial support over very short terms will be unlikely to be successful. Examples of the effective use of donor funds have been shown for projects that provide specific support to private-sector businesses. In particular, social enterprises and NGOs help create sustainable markets, such as by using targeted subsidies and incentives to encourage the delivery of energy to rural areas or poor urban areas.

The roles of formal private-sector financing institutions in delivering energy to the poor is still not well developed, although social impact investors have started to work in this area with angel investments being provided together with grants, and/or subsidies from governments or the donor community. The carbon credit market has also started to provide some finance for energy delivery models, although it has been subject to great instability in recent years. It does not offer a solid and secure base for long-term investment in energy delivery models, although in a few instances it has helped off-grid energy technologies get off the ground in some developing countries. It is important to note that as carbon finance is effectively a short-term subsidy, it has the potential to distort local markets and affect the long-term sustainability of energy delivery models.

Capacity building

Capacity building is often used as a catch-all term for a range of activities and services, but is an essential requirement for the effective and sustainable delivery of energy services to the poor. Specific capacity-building services are required for the energy market chain actors, such as energy resource data; support for small, informal entrepreneurs; as well as for the end users themselves. Capacity building needs to cover technical and managements competencies, which are often scarce or disproportionally expensive, particularly in remote or rural areas. Capacity building is often required right across all scales of energy delivery models. Support should be provided to on-grid energy utilities to ensure they can design systems that meet the needs of the poor; and to mini-grid management companies that ensure community participation and buy-in is maximized and that the systems are sustainably designed.

Capacity building is also required for the companies designing and manufacturing the energy technologies, as well as for their ongoing maintenance. It is also required for the end users to ensure they understand the performance and limitations of the technologies, as well as their affordability. Effective awareness-raising for the end user is often one of the most significant barriers to increasing energy access for the poor; by improving the understanding of how the technologies perform and what level of energy service each solution can achieve, false expectations and potential rejection or abandoning of systems by the end user can be avoided.

Capacity building is required for some of the other supporting services, as well as the enabling environment actors: for the energy companies to understand how to use financial services more effectively; for financial institutions which often have a very poor understanding of how energy technologies or products work, especially in rural areas, and are thereby unable to structure their financing products accurately; and for a range of national and local government departments to better understand the full range of energy services that are needed to overcome poverty, including the technologies, costs, design, and barriers to up-scaling.

Enabling environment

As highlighted by a number of case studies in this book, the environment that surrounds energy delivery models with development goals is crucial to their success. Regulations, national policies, the drafting of standards, and the creation of quality control bodies, as well as the appreciation of the local socio-cultural context, are essential to ensuring that energy services can be effectively and sustainably delivered.

National policies can greatly influence the attractiveness of a country for investors. An obvious example is the use of subsidies which can help with the uptake of new, sustainable energy solutions within a market in the short term, or create perverse incentives, such as is the case with many fossil fuel-based energy services in developing countries. The countries that have committed to achieving sustainable energy access for all in the relatively near future are currently developing a range of new measures. These include bold national and regional energy policy targets; the developing of appropriate tariff structures; the supporting of financial incentives to welcome private-sector participation; and the increasing of programmatic capacity. In drawing up their energy strategies, an increasing number of countries are beginning to understand the development benefits of energy access, and that the eradication of energy poverty is a fundamental development goal without which many of the other development goals will be impossible to achieve. Clear targets to reach the poorest and most isolated citizens have highlighted the importance of considering all energy needs, including household cooking and the powering of productive uses for income generation and increased agricultural production. Also the dominance of on-grid energy delivery within many countries' national energy strategies is beginning to decrease, with the prominence of mini-grid and off-grid systems starting to rise, both for electrification and cooking.

The challenges faced in supporting the implementation of development projects still require a range of measures for success which combine supporting services, favourable policy environments, and a multitude of different partners. Collaborations have stretched through a range of actors, from well-coordinated government departments, to donors, financial institutions, and private-sector companies, including social enterprises, civil society, and capacity-building institutions (e.g. universities). Concerted action will make feasible the eradication of energy poverty and the achievement of universal energy access.

References

Advisory Group on Energy and Climate Change (AGECC) (2010) *Energy for a Sustainable Future, Report and Recommendations,* New York: UN AGECC <www.un.org/wcm/webdav/site/climatechange/shared/Documents/ AGECC%20summary%20report%5B1%5D.pdf> [accessed June 2013].

Agbaje, T. (2009) 'From user involvement to user initiative: the role of priority identification in facilitating sustainability of rural renewable energy projects', *EWB-UK Research Conference,* 20 February <www.ewb-uk.org/ filestore/Full%20Proceedings%202009.pdf> [accessed 22 July 2013].

Agbemabiese, L. (2008) 'Expanding energy access through sustainable energy enterprises in Africa: financing, capacity-building and policy aspects', in *Energy Poverty in Africa: Proceedings of a Workshop held by OFID in Abuja, Nigeria, June 8–10, 2008,* pp. 121–138, OFID Pamphlet Series 39, Vienna, Austria: OPEC Fund for International Development <www.ofid.org/ LinkClick.aspx?fileticket=xNKy_XeYw7g%3D&tabid=109> [accessed 27 June 2013].

Ahmad, B. (2001) 'Users and disusers of box solar cookers in urban India: implications for solar cooking projects', *Solar Energy* 69 (Supp. 6): 209–15 <http://dx.doi.org/10.1016/S0038-092X(01)00037-8>

Albu, M. and Griffith, A. (2005) 'Mapping the market: a framework for rural enterprise development policy and practice', Rugby, UK: Practical Action <http:// practicalaction.org/docs/ia2/mapping_the_market.pdf> [accessed 1 May 2013].

Alliance for Rural Electrification (ARE) (2011) *Hybrid Mini-Grids for Rural Electrification: Lessons Learned,* Brussels: ARE-USAID <www.ruralelec.org/ fileadmin/DATA/Documents/06_Publications/Position_papers/ARE_Mini-grids_-_Full_version.pdf> [accessed 27 June 2013].

Amarasekara, R.M. and Atukorala, K. (no date) *Historical Timeline for Subsidy to Commercialisation of Improved Cookstoves: Path Leading to Sustainable Stove Development and Commercialisation in Sri Lanka,* Sri Lanka: Integrated Development Association (IDEA) <www.ideasrilanka.org/PDFDownloads/Historical%20 timelines%20of%20stoves%20sri%20lanka.pdf> [accessd 27 June 2013].

Aron, J.E., Kayser, O., Liautaud, L., and Nowlan, A. (2009) *Access to Energy for the Base of the Pyramid,* Paris: Hystra, and Arlington, VA: Ashoka <http://hystra. com/energy> [accessed 27 August 2013].

Arze del Granado, J., Coady, D. and Gillingham, R. (2010) *The Unequal Benefits of Fuel Subsidies: A Review of Evidence for Developing Countries, International Monetary Fund (IMF) Working Paper No. WP/10/202,* Washington: IMF <www. imf.org/external/pubs/ft/wp/2010/wp10202.pdf> [accessed 1 May 2013].

Asfaw, A. and Demissie, Y. (2012) 'Sustainable household energy for Addis Ababa, Ethiopia', *Consilience: The Journal of Sustainable Development,* 8: 1–11 <www.consiliencejournal.org/index.php/consilience/article/viewFile/252/110>.

Ashden Awards (2011) 'Case study summary: Husk Power Systems India', London: Ashden <www.ashden.org/files/Husk%20winner.pdf> [accessed 27 June 2013].

Aylett, A. (2010) 'Solar sisters: the Avon lady of African renewables' [blog], 2 June, *World Changing* <www.worldchanging.com/archives/011236.html> [accessed on 2 May 2013].

Barnes, D.F, Openshaw, K., Smith, K.R and van der Plas, R. (1994) *What Makes People Cook with Improved Biomass Stoves?: A Comparative Review of Stove Programs, World Bank Technical Paper No. 242*, Washington, DC: The World Bank <http://ehs.sph.berkeley.edu/krsmith/publications/94_barnes_1.pdf> [accessed 27 June 2013].

Barnes, D.F (2005) *Meeting the Challenge of Rural Electrification in Developing Nations: The Experience of Successful Programs*, Washington, DC: Energy Sector Management Assistance Program (ESMAP), World Bank<http://siteresources.worldbank.org/EXTRENENERGYTK/Resources/51+38246-1237906527727/5950705-1239305592740/Meeting0the0Ch10Discussion0Version0.pdf> [accessed 27 June 2013].

Barnes, D., Krutilla, K., and Hyde, W. (2005) *The Urban Household Energy Transition*, Washington. DC: Resources for the Future Press.

Bazilian, M., Nussbaumer, P., Cabraal, A., Centurelli, R., Detchon, R., Geilen, D., Rogner, H., Howells, M., McMahon, H., Modi, V., Nakicenovic, N., O'Gallochoir, B., Radka, M., Rijal, K., Takada, M. and Ziegler, F. (2010) 'Measuring energy access: supporting a global target', New York, NY: Earth Institute, Columbia University<www.unido.org/fileadmin/user_media/Services/Energy_and_Climate_Change/EPP/Publications/bazilian%20et%20al%202010%20measuring%20energy%20access%20supporting%20a%20global%20target.pdf> [accessed 27 June 2013].

Behrens, A., Lahn, G., Dreblow, E., Ferrer, J.N, Carraro, M. and Veit, S. (2012) *Escaping the Vicious Cycle of Poverty: Towards Universal Access to Energy*, CEPS Working Document, No. 363, Brussels: Centre for European Policy Studies (CEPS) <http://aei.pitt.edu/33836/> [accessed 25 April 2013].

Bellanca, R. (2012a) 'Interview with Seena Rejal, Eight19 Limited', HEDON <www.hedon.info/IIED+SE4All+Interview_Eight19+SRejal> [accessed 27 June 2013].

Bellanca, R. (2012b) 'Interview with Harry Stokes, Project Gaia Inc.', HEDON <www.hedon.info/IIED+SE4All+Interview_ProjectGaiaInc+HStokes?bl=y> [accessed 27 June 2013].

Bellanca, R. (2012c) 'Interview with Michael Kelly, World LP Gas Association', HEDON <www.hedon.info/IIED+SE4All+Interview_WLPA+MKelly?bl=y> [accessed 27 June 2013].

Bellanca, R. (2012d) 'Interview with Sarah Alexander, SELCO India', HEDON <www.hedon.info/IIED+SE4All+Interview_SELCO+SAlexander> [accessed 27 August 2013].

Bellanca, R. (2012e) 'Interview with Sagun Saxena and Azzurra Massimino, CleanStar Ventures LLC', HEDON <www.hedon.info/IIED+SE4All+-Interview_CleanStar+SSaxena> [accessed 27 August 2013].

Bellanca, R. and Garside, B. (forthcoming) *Designing Energy Delivery Models that Work for Poor People*, London: International Institute for Environment and Development (IIED).

Bellanca, R. and Wilson, E. (2012) 'Sustainable Energy for All and the private sector', London: IIED <http://pubs.iied.org/G03383.html> [accessed 27 June 2013].

Best, S. (2011) *Remote Access: Expanding Energy Provision in Rural Argentina through Public–Private Partnerships and Renewable Energy: A Case Study of the*

PERMER Programme, London: IIED <http://pubs.iied.org/16025IIED.html> [accessed 2 May 2013].

Biolite (no date) 'Biolite Homestove: bringing clean energy to families around the world' <www.biolitestove.com/> [accessed 2 May 2013].

Bloomfield, E. (2012) 'Bioenergy market system development: comparing participatory approaches in Kenya and Sri Lanka', *Boiling Point* 60: 6–9 <www.hedon.info/View+issue&itemId=12257> [accessed 25 April 2013].

Bruce, N., Perez-Padilla, R. and Alablak, R. (2000) 'Indoor air pollution in developing countries: a major environmental and public health challenge', *WHO Bulletin 2000* 78: 1078–92.

Cecelski, E. (2004) *Re-thinking Gender and Energy: Old and New Directions, ENERGIA/EASE Discussion Paper* <www.africaadapt.net/media/resources/65/ENERGIA.pdf> [accessed 25 April 2013].

Cherni, J.A and Preston, F. (2007) 'Rural electrification under liberal reforms: the case of Peru', *Journal of Cleaner Production* 15: 143–52.

Clancy, J., Skutch, M. and Batchelor, S. (2004) *The Gender-Energy-Poverty Nexus: Finding the Energy to Address Gender Concerns in Development,* London: Department for International Development (DFID) <www.esmap.org/sites/esmap.org/files/The%20Gender%20Energy%20Poverty%20Nexus.pdf> [accessed 27 June 2013].

CleanStar Mozambique (no date) [website] <www.cleanstarmozambique.com/> [accessed 27 June 2013].

Cotula, C., Finnegan, L. and Macqueen, D. (2011) *Biomass Energy: Another Driver of Land Acquisitions?*, London: IIED <http://sd-cite.iisd.org/cgi-bin/koha/opac-detail.pl?biblionumber=51869> [accessed 25 April 2013].

CRELUZ (no date) 'Micro-hydro makes the grid reliable' <www.ashden.org/winners/CRELUZ10> [accessed 27 June 2013].

de Gouvello, C. and Kumar, G. (2007) 'Output-based aid in Senegal – designing technology-neutral concessions for rural electrification', *OBA Approaches*, Note No. 14, Washington, DC: Global Partnership on Output-based Aid (GPOBA), World Bank <www.gpoba.org/sites/gpoba.org/files/OBApproaches14_SenegalElectric.pdf> [accessed 2 May 2013].

Deininger, K., Byerlee, D., Lindsay, J., Norton, A., Selod, H., and Stickler, M. (2011) *Rising Global Interest in Farmland: Can it Yield Sustainable and Equitable Benefits?*, Washington, DC: World Bank <http://siteresources.worldbank.org/INTARD/Resources/ESW_Sept7_final_final.pdf [accessed 1 May 2013].

Department of Energy (2009) 'Socio-economic impact of electrification 2008 and 2009', Pretoria, South Africa: Department of Energy.

Development and Energy in Africa (DEA) (2011) 'Development and Energy in Africa (DEA) project: a case for Botswana', Botswana National Paper, DEA <http://ebookbrowse.com/botswana-national-background-paper-doc-d420571560> [accessed 25 April 2013].

Department for International Development (DFID) (2002) *Energy for the Poor: Underpinning the Millennium Development Goals*, London: Future Energy Solutions / DFID <www.ecn.nl/fileadmin/ecn/units/bs/JEPP/energyforthepoor.pdf> [accessed 25 April 2013].

Energy Information Administration (EIA) (2005) 'International energy statistics', Washington, DC: US Department of Energy <www.eia.gov/cfapps/ipdbproject/IEDIndex3.cfm?tid=2&pid=2&aid=12> [accessed 29 August 2013].

Energy Sector Management Assistance Programme (ESMAP) (2000) *Reducing the Cost of Grid Extension for Rural Electrification*, Washington, DC: ESMAP <http://rru.worldbank.org/Documents/PapersLinks/1072.pdf> [accessed 1 May 2013].

ESMAP (2005) *Transformative Power: Meeting the Challenge of Rural Electrification in Developing Nations: The Experience of Successful Programs, Knowledge Exchange Series, No. 2*, Washington, DC: ESMAP <www.esmap.org/sites/esmap.org/files/KES02_Transformative%20Power%20Meeting%20the%20Challenge%20of%20Rural%20Electrification.pdf> [accessed 27 June 2013].

ESMAP (2008) *Regulatory Review of Power Purchase Agreements: A Proposed Benchmarking Methodology, Report No. 337/08*, Washington, DC: ESMAP <http://siteresources.worldbank.org/EXTAFRREGTOPENERGY/Resources/717305-1266613906108/ESMAP_337-08_Regulatory.pdf> [accessed 27 June 2013].

EPA (United States Environmental Protection Agency) (2013) 'Climate change: global greenhouse gas emissions data: emissions by country', Washington, DC: EPA <www.epa.gov/climatechange/ghgemissions/global.html#four> [accessed 26 April 2013].

EU Energy Initiative – Partnership Dialogue Facility (EUEI-PDF) (2008) *Development of an Energy Access Strategy: Scaling up Access to Modern Services in the East African Community: A Policy Brief*, EUEI-PDF: Germany <www.euei-pdf.org/regional-studies/development-of-an-energy-access-strategy> [accessed 26 April 2013].

Ferrer-Martí, L., Garwood, A., Chiroque, J., Ramirez, B., Marcelo, O., Garfi, M. and Velo, E. (2012) 'Evaluating and comparing three community small-scale wind generation projects', *Renewable and Sustainable Energy Reviews* 16: 5379–90.

Foster, V., Tre, J. P. and Wodon, Q. (2000) *Energy Prices, Energy Efficiency, and Fuel Poverty*, Washington, DC: World Bank <http://info.worldbank.org/etools/docs/voddocs/240/502/Gua_price.pdf> [accessed 26 April 2013].

FAOSTAT-ForesSTAT (2011) 'Highlights on wood charcoal: 2004–2009', Rome: Food and Agriculture Organization (FAO) <http://faostat.fao.org/Portals/_Faostat/documents/pdf/Wood%20charcoal.pdf> [accessed 26 April 2013].

Global Environment Facility (GEF) (2009) *Investing in Renewable Energy: the GEF Experience*, Washington, DC: GEF.

Global Network on Sustainable Development (GNESD) (2007) *Reaching the Millennium Development Goals and Beyond: Access to Modern Forms of Energy as a Prerequisite*, Nairobi: United Nations Environment Programme (UNEP) <www.gnesd.org/upload/gnesd/pdfs/other%20publications/mdg_energy.pdf> [accessed 27 June 2013].

Global Subsidies Initiative (GSI) (2011) 'A high-impact initiative for Rio+20: a pledge to phase out fossil-fuel subsidies', Geneva: The International Institute for Sustainable Development <www.iisd.org/pdf/2011/joint_ngo_submission_rio_plus_20.pdf> [accessed 1 May 2013].

Global Village Energy Partnership (GVEP) International (no date) *Developing Energy Enterprises in East Africa: Summary Report*, London: GVEP International, <www.gvepinternational.org/sites/default/files/deep_booklet_2013_0.pdf> [accessed 27 June 2013].

GVEP International (2008) *International GAP Fund Report*, London: GVEP International.

GVEP International (2010) *Training Manual for Senior and Middle Level Managers in Energy Financing*, London: GVEP International <www.gvepinternational. org/sites/default/files/manual_for_senior_managers_of_fis.pdf>[last accessed 27 June 2013].

GVEP International (2011) 'The history of mini-grid development in developing countries', *Policy Briefing*, September, London: GVEP International <www. gvepinternational.org/sites/default/files/policy_briefing_-_mini-grid_final. pdf> [last accessed 27 June 2013].

GVEP International (2012) 'How loan guarantees can stimulate small energy enterprises', London: GVEP International <www.gvepinternational.org/en/ business/access-finance> [accessed 2 May 2013].

Global Wind Energy Council (GWEC) (no date), [website] <www.gwec.net> [accessed 2 May 2013].

Gold Standard (no date) 'Clean cooking in conflict-hit Darfur', Geneva: The Gold Standard Foundation <www.cdmgoldstandard.org/clean-cooking-in-conflict-hit-darfur> [accessed 2 May 2013].

Graham, F. (2010) 'M-Pesa: Kenya's mobile wallet revolution', *BBC News* <www. bbc.co.uk/news/business-11793290> [accessed 26 June 2013].

Grameen Shakti (no date) [website] <www.gshakti.org> [accessed 27 June 2013].

Greacen, C.S and Greacen, C. (2004) 'Thailand's electricity reforms: privatization of benefits and socialization of costs and risks', *Pacific Affairs* 77: 517–41.

Greacen, C. (2007) 'An emerging light: Thailand gives the go-ahead to distributed energy', *Cogeneration & On-Site Power Production Magazine* [online] <www.cospp.com/articles/print/volume-8/issue-2/features/an-emerging-light-thailand-gives-the-go-ahead-to-distributed-energy.html> [accessed 24 July 2013].

Group for the Environment, Renewable Energy and Solidarity (GERES) (2009) *Dissemination of Domestic Efficient Cookstoves in Cambodia: Looking Back on a 10-year Programme Combining Development, Fight against Climate Change and Environmental Protection*, Aubagne, France: Association GERES <www. geres.eu/en/resources/publications/item/160-dissemination-of-domestic-efficient-cookstoves-in-cambodia> [accessed 26 April 2013].

Guerra-Garcia, E. (2004) 'La sociointerculturalidad y la educación indígena', in E.A. Sandoval Forero and M.A. Baeza (eds), *Cuestión étnica, culturas, construcción de identidades*, Mexico: UAIM, ALAS, El Caracol.

Haanyika, C.M (2006) 'Rural electrification policy and institutional linkages', *Energy Policy* 34: 2977–93 <http://dx.doi.org/10.1016/j.enpol.2005.05.008>.

Hamilton, K. (2010) *Scaling Up Renewable Energy in Developing Countries: Finance and Investment Perspectives*, Energy, Environment & Resource Governance Programme Paper 02/10, London: Chatham House <www.chathamhouse. org/sites/default/files/public/Research/Energy,%20Environment%20and%20 Development/0410pp_hamilton.pdf> [accessed 27 June 2013].

Hammond, A.L., Kramer, W.J., Katz, R.S., Tran, J.T. and Walker, C. (2007) 'The energy market', in *The Next 4 Billion: Market Size and Business Strategy at the Base of the Pyramid*, pp. 76–87, Washington, DC: World Resources Institute and International Finance Corporation <http://pdf.wri.org/n4b_chapter7. pdf> [accessed 24 July 2013].

Heimann, R.P., Parkinson, E., Vullioud, G.A, Keck, H., Hauser, H.P., Keiser, W. and Rothenfluh, M. (2008) *Systematic Approach of Pelton Rehabilitation Projects: Practical Experience from Case Studies, Proceedings of HydroVision 2008*, Florida: HCI Publications.

Intergovernmental Panel on Climate Change (IPCC) (2011) *IPCC Special Report on Renewable Energy Sources and Climate Change Mitigation*, prepared by Working Group III of the Intergovernmental Panel on Climate Change, Cambridge and New York, NY: Cambridge University Press.

International Energy Agency (IEA) (2011) *2011 Key World Energy Statistics*, Paris: IEA <www.iea.org/publications/freepublications/publication/key_world_energy_stats-1.pdf>.

IEA Bioenergy (2009) *Bioenergy – A Sustainable and Reliable Energy Source: A Review of Status and Prospects*, Rotorua, New Zealand: IEA Bioenergy <www.ieabioenergy.com/libitem.aspx?id=6479> [accessed 27 June 2013].

IEA Geothermal (2012) *Trends in Geothermal Applications: Survey Report on Geothermal Utilization and Development in IEA-GIA Member Countries in 2010, with Trends in Geothermal Power Generation and Heat Use 2000–2010*, Tuapo, New Zealand: IEA-GIA <http://iea-gia.org/wp-content/uploads/2012/08/GIA_TrendsGeothermalApplications-2010_Vs2_1-Ganz-29Aug12.pdf> [accessed 2 May 2013].

International Finance Corporation (IFC) (2012) *Lighting Asia: Solar Off-Grid Lighting: Market Analysis of India, Bangladesh, Nepal, Pakistan, Indonesia, Cambodia and Philippines*, New Delhi: IFC <www.ifc.org/wps/wcm/connect/topics_ext_content/ifc_external_corporate_site/ifc+sustainability/publications/publications_report_lightingasia> [accessed June 2013].

International Hydropower Association (IHA) (2011) *Advancing Sustainable Hydropower: 2011 Activity Report*, London: IHA <www.hydropower.org/downloads/ActivityReports/2011-12_Activity_Report-web.pdf> [accessed 27 June 2013].

International Renewable Energy Agency (IRENA) (2011) *Prospects for the African Power Sector: Scenarios and Strategies for Africa project*, Abu Dhabi: IRENA <www.irena.org/DocumentDownloads/Publications/Prospects_for_the_African_PowerSector.pdf> [accessed 27 June 2013].

IRENA (2012a) *Hydropower,* Working Paper, Renewable Energy Technologies: Cost Analysis Series, Vol. 1: Power Sector, No. 3, Abu Dhabi: IRENA <www.irena.org/DocumentDownloads/Publications/RE_Technologies_Cost_Analysis-HYDROPOWER.pdf> [accessed 2 May 2013].

IRENA (2012b) *Concentrating Solar Power*, Working Paper, Renewable Energy Technologies: Cost Analysis Series, Vol. 1: Power Sector, No. 2), Abu Dhabi: IRENA <www.irena.org/DocumentDownloads/Publications/RE_Technologies_Cost_Analysis-CSP.pdf> [accessed 2 May 2013].

International Rivers (no date) 'Grand Inga Dam, DR Congo' [online], Berkeley, CA: International Rivers <www.internationalrivers.org/campaigns/grand-inga-dam-dr-congo> [accessed 2 May 2013].

International Science Panel on Renewable Energies (ISPRE) (2009) *Research and Development on Renewable Energies: A Global Report on Photovoltaic and Wind Energy*, Paris: ISPRE<www.icsu.org/publications/reports-and-reviews/ispre-photovoltaic-wind/ISPRE_Photovoltaic_and_Wind.pdf> [accessed 27 June 2013].

Jacobson, A. (2007) 'Connective power: solar electrification and social change in Kenya', *World Development* 35: 144–62.

Jiahua, P., Meng, L., Meng, L., Xiangyang, W., Lishuang, W., Elias, R.J., Victor, D.G., Zerriffi, H., Zhang, C. and Wuyuan, P. (2006), *Rural Electrification in China 1950–2004: Historical Processes and Key Driving Forces*, Working Paper Number 60, Stanford, CA: Program on Energy and Sustainable Development, Stanford University.

Johansson, O., Lauenburg, P. and Wollerstrand, J. (2010) 'District heating in case of power failure', *Applied Energy* 87: 1176–86.

Johnson, F.X. and Lambe, F. (2009) 'Energy access, climate and development', Stockholm: Commission on Climate Change and Development <www.environmentportal.in/files/ccd_energyaccessclimateanddev2009.pdf> [accessed 26 April 2013].

Karakezi, S. and Walubengo, D. (1989) *Household Stoves in Kenya: The Case of the Kenya Ceramic Jiko*, Nairobi: Kenya Energy and Environment Organization.

Kariuki, F. and Rai, K. (2010) *Market Survey on Possible Co-operation with Finance Institutions for Energy Financing in Kenya, Uganda and Tanzania*, London: GVEP International   <www.gvepinternational.org/sites/default/files/financial_institutions_market_study_in_east_africa_2010_gvep_international.pdf> [accessed 27 June 2013].

Kartha, S. and Larson, E.D. (2000) *Bioenergy Primer: Modernised Biomass Energy for Sustainable Development*, New York, NY: United Nations Development Programme <www.undp.org/content/undp/en/home/librarypage/environment-energy/sustainable_energy/bioenergy_primermodernisedbiomassenergyforsustainabledevelopment/> [accessed 27 August 2013].

Khennas, S. and Barnett, A. (2000) *Best Practices for Sustainable Development of Micro Hydro Power in Developing Countries*, London: DFID <www.microhydropower.net/download/bestpractsynthe.pdf> [accessed 27 June 2013].

Knight, O. (2011) 'Results-based financing', presentation to SREP sub-committee, Cape Town, South Africa, 21 June. Washington, DC: ESMAP, World Bank <www.esmap.org/sites/esmap.org/files/Results-Based%20Financing%20-%20SREP%20%5BSub-Committee%20Meeting%5D%20-%20Oliver%20Knight.pdf> [accessed 2 May 2013].

Krishnaswamy, S. (2010) *Shifting of Goal Posts – Rural Electrification in India: A Progress Report*, New Delhi: Vasudha Foundation <www.christianaid.org.uk/images/shifting-goal-posts.pdf> [accessed 27 June 2013].

Lighting Africa (no date) 'Market intelligence' [online] <www.lightingafrica.org/what-we-do/market-intelligence.html> [accessed 2 May 2013].

Lighting Africa (2012) 'Policy report note: Ghana', Washington, DC: International Finance Corporation and World Bank <www.ana.lightingafrica.org/index.php/component/docman/doc_details/306-ghana-policy-report-note.html> [accessed 27 June 2013].

Macqueen, D. and Korhaliller, S. (2011) *Bundles of Energy: The Case for Renewable Biomass Energy*, London: IIED <http://pubs.iied.org/13556IIED.html> [accessed 2 May 2013].

Martinot, E., Cabraal, A. and Mathur, S. (2000) 'World Bank solar home systems projects: experiences and lessons learned 1993–2000', *Renewable and Sustainable Energy Reviews*, 5: 39–57.

Mitchell et al. (2011) 'Policy, Financing and Implementation', in: *IPCC Special Report on Renewable Energy Sources and Climate Change Mitigation*, p 865–950, Cambridge and New York: Cambridge University Press.

Mketsi, M. (2009) Director Electricity Supply, Department of Energy, South Africa. Interview by H.R. Gron, May.

Modi, V., McDade, S., Lallement, D., and Saghir, J. (2006) *Energy and the Millennium Development Goals*, New York, NY: Energy Sector Management Assistance Programme, UN Development Programme and World Bank <www.unmillenniumproject.org/documents/MP_Energy_Low_Res.pdf> [accessed 27 June 2013].

Morris, E., Winiecki, J., Chowdhary, S. and Cortiglia, K. (2007) *Using Microfinance to Expand Access to Energy Services: Summary of Findings*, Washington, DC: SEEP Network <www.seepnetwork.org/using-microfinance-to-expand-access-to-energy-services--summary-of-findings-resources-598.php> [accessed 27 June 2013].

Motta, M. and Reiche, K. (2001) *Rural Electrification, Micro-finance and Micro and Small Business (MSB) Development: Lessons for the Nicaragua Off-Grid Rural Electrification Project*, Washington, DC: World Bank <http://siteresources.worldbank.org/EXTRENENERGYTK/Resources/5138246-1237906527727/5950705-1239305592740/RuralelectrificationMicrofinance.pdf> [accessed 2 May 2013].

Mumssen, Y., Johannes, L. and Kumar, G. (2010) *Output-Based Aid: Lessons Learned and Best Practices*, Washington, DC: World Bank.

Munasinghe, M. (1987) *Rural Electrification for Development: Policy Analysis and Applications*, Boulder, CO: Westview Press.

Mureithi, F. (2012) 'Kenya Power's "Stima Loan" in Sh3bn boost', *Sunday Nation*, 15 September <www.nation.co.ke/business/news/Kenya+Powers+Stima+loan+boost/-/1006/1508396/-/qffavf/-/index.html> [accessed 27 June 2013].

Namy, S. (2007) 'Addressing the social impacts of large hydropower dams', *The Journal of International Policy Solutions* 7: 11–17 <http://irps.ucsd.edu/assets/012/6359.pdf>.

Navas-Sabater, J., Dymond, A. and Juntunen, N. (2002) *Telecommunications and Information Services for the Poor: Toward a Strategy for Universal Access*, World Bank Discussion Paper No. 432, Washington, DC: World Bank.

Nexus-Carbon for Development (no date) [website] <www.nexus-c4d.org> [accessed 2 May 2013].

Nigeria LNG (no date) 'Bonny Utility Company' <www.nlng.com/PageEngine.aspx?&id=22> [accessed 27 June 2013].

Njuguna, J. (2012) 'Electrification strategies "Stima Loan" facilities, Kenya Power and Lighting Company (KPLC) presentation', *ESMAP-Cities Alliance Workshop*, Washington, DC, May.

NRECA International (2010) *Guides for Electric Cooperative Development and Rural Electrification*, Arlington, VA: NRECA International <www.nrecainternational.coop/resources/Publications/Pages/GuidesforElectricCooperativeDevelopmentandRuralElectrification.aspx>.

Nyabundi, D. (2012) 'Stima Loan helps poor families get electricity', *Sunday Nation*, 4 December <www.businessdailyafrica.com/Corporate-News/Stima-Loan-helps-poor-families-get-electricity/-/539550/1636790/-/kkwkoyz/-/index.html> [accessed 27 June 2013].

Nygaard, I. (2009) 'The compatibility of rural electrification and promotion of low-carbon technologies in developing countries: the case of solar PV for sub-Saharan Africa', *European Review of Energy Markets* 3: 1–34.

Organisation for Economic Co-operation and Development / International Energy Agency (OECD/IEA) (2004) *World Energy Outlook 2004*, Paris: IEA

<www.oecd-ilibrary.org/energy/world-energy-outlook-2004_weo-2004-en> [accessed 27 June 2013].

OECD/IEA (2006) *World Energy Outlook 2006*, Paris: IEA <www.oecd-ilibrary.org/energy/world-energy-outlook-2006_weo-2006-en> [accessed 27 June 2013].

OECD/IEA (2010) *World Energy Outlook 2010*, Paris: IEA <www.oecd-ilibrary.org/energy/world-energy-outlook-2010_weo-2010-en> [accessed 27 June 2013].

OECD/IEA (2011) *World Energy Outlook 2011*, Paris: IEA <www.oecd-ilibrary.org/energy/world-energy-outlook-2011_weo-2011-en> [accessed 27 June 2013].

OECD/IEA (2012) *World Energy Outlook 2012,* Paris: IEA <www.oecd-ilibrary.org/energy/world-energy-outlook-2012_weo-2012-en> [accessed 27 June 2013].

Perlack, R.D., Jones, H.G. and Waddle, D.B. (1990) 'A survey of renewable energy technologies for rural applications', *Energy* 15: 1119–27.

PISCES (no date) [website] <www.pisces.or.ke/node/1> [accessed 2 May 2013].

Polak, P. (2003) 'How IDE installed 1.3 million treadle pumps in Bangladesh by activating the private sector: the practical steps', Berne: Swiss Agency for Development and Co-operation <www.its.caltech.edu/~e105/readings/cases/IDEtreadle.pdf> [accessed 27 June 2013].

PowerPot (no date) [website] <www.thepowerpot.com/> [accessed 2 May 2013].

Practical Action (no date) 'Energy delivery model tool' [online] <http://practicalaction.org/consulting/pisces/> [accessed 2 May 2013].

Practical Action (2010) *Poor People's Energy Outlook 2010*, Rugby, UK: Practical Action Publishing <http://practicalaction.org/docs/energy/poor-peoples-energy-outlook.pdf> [accessed 27 June 2013].

Practical Action (2012) *Poor People's Energy Outlook 2012*, Rugby, UK: Practical Action Publishing <http://practicalaction.org/ppeo2012> [accessed 27 June 2013].

Practical Action Consulting (2009a) 'Scale-up and commercialisation of improved cookstoves in Sri Lanka: the Anagi experience, Working Paper, London: PISCES <http://r4d.dfid.gov.uk/Output/182835/Default.aspx> [accessed 27 June 2013].

Practical Action Consulting (2009b) *Small-Scale Bioenergy Initiatives: Brief Description and Preliminary Lessons on Livelihoods, Impacts from Case Studies in Asia, Latin America and Africa*, Rome: FAO; and Nairobi: PISCES <ftp://ftp.fao.org/docrep/fao/011/aj991e/aj991e.pdf> [accessed 27 June 2013].

Practical Action Consulting (2011) 'Energy delivery model tool for understanding and scaling up decentralised energy supply', in *Proceedings of the International Conference on Micro Perspectives for Decentralised Energy Supply*, pp. 280–1, Technische Universitat Berlin, Universitätsbibliothek, Berlin, 7–8 April.

Practical Action Consulting (2012) *Biomass Gasification: The East African Study*, Working Paper, London: PISCES <www.pisces.or.ke/sites/default/files/04398%20PAC%20Biomass%20Gasification%20LR.pdf> [accessed 27 June 2013].

Prasad, D. (2013) Direct correspondence with Mr Dilli Prasad Ghimire, Chairman, National Association of Community Electricity Users-Nepal, February.

Qiu, D. and Gu, S. (1996) 'Diffusion of improved biomass cookstoves in China', *Energy Policy* 24: 463–69 <http://dx.doi.org/10.1016/0301-4215(96)00004-3>.

Rai, K. and McDonald, J. (eds) (2009) 'The commercialisation and scale up success of improved cookstoves in Sri Lanka', *Cookstoves and Markets: Experiences, Successes and Opportunities,* London: GVEP International <www.hedon.info/docs/GVEP_Markets_and_Cookstoves__.pdf> [accessed 26 April 2013].

Reddy, A.K.N. (2004) 'Lessons from the Pura community biogas project', *Energy for Sustainable Development* 8 (3): 42.

Reegle (2006) 'Energy profile Laos' [online] <www.reegle.info/countries/laos-energy> [accessed 27 June 2013].

Reegle (2011) 'Costa Rica 2011' [online] <www.reegle.info/policy-and-regulatory-overviews/CR> [accessed 2 May 2013].

Reegle (2012) 'Bangladesh 2012' [online] <www.reegle.info/policy-and-regulatory-overviews/BD> [accessed 2 May 2012].

Reiche, K., Covarrubias, A. and Martinot, E. (2000) 'Expanding electricity access to remote areas: off-grid rural electrification in developing countries', *WorldPower* (2000): 52–60 <http://w.martinot.info/Reiche_et_al_WP2000.pdf> [accessed 29 April 2013].

Renewable Energy Policy Network for the 21st Century (REN21) (2011) *Renewables 2011: Global Status Report*, Paris: REN21 <www.ren21.net/Portals/0/documents/Resources/GSR2011_FINAL.pdf> [accessed 27 June 2013].

Renewable World (2012) *Making a Unique and Meaningful Contribution to Clean Energy Access for the Poorest on the Planet*, Brighton: Renewable World <www.renewable-world.org/sites/default/files/RW%20position_paper_final%20LOW%20RES.pdf> [accessed 27 June 2013].

Rufin, C. and Márquez, P. (2011) *Private Utilities and Poverty Alleviation: Market Initiatives at the Base of the Pyramid*, Cheltenham, UK: Edward Elgar Publishing.

Rural Energy for Rural Economic Development (RERED) (2012) 'RERED project comes to a successful closure' [online] <www.energyservices.lk/pdf/rered_%20projects_%20comes_to_successful_closure.pdf> [accessed 2 May, 2013].

Sanchez, T. (2006) *Electricity Services in Remote Rural Communities: The Small Enterprise Model*, Rugby, UK: Practical Action Publishing.

Sanchez, T. (2008) *Delivery Models for Energy Access for the Poor*, PISCES Scoping Study, PISCES Research Programme Consortium (RPC) / Practical Action Consulting (unpublished).

Sanchez, T. (2010) *The Hidden Energy Crisis: How Policies are Failing the World's Poor*, Rugby, UK: Practical Action Publishing.

Schneider Electric (2012) *Share: 2011–2012 Strategy and Sustainability Highlights*, France: Schneider Electric <www.schneider-electric.com/documents/interactive-publications/20112012_sustainable_and_development_highlights_in_brief/files/docs/all.pdf> [accessed 27 June 2013].

Schützeichel, H. (2012) *Ethiopia Solar: The Initiation of a Solar Trade in Ethiopia 2005–2011*, Merzhausen/Zurich/Addis Ababa: Solar Energy Foundation <www.solar-energy-foundation.nl/pdf-files/Ethiopia%20solar_GB_final.pdf> [accessed 27 June 2013]

SELCO Solar (no date) 'Access to energy services: 8 case studies', Bangalore: SELCO, and Vienna: REEEP <www.selco-india.com/pdfs/case_stories2.pdf> [accessed 27 August 2013].

Sinton, J.E, Smith, K.R, Peabody, J.W, Liu, Y., Zhang, X. and Edwards, R. (2004) 'An assessment of programs to promote improved household stoves in China', *Energy Sustainable Development* 3:33–52.

Stevens, L. (2010) 'Practical Action news: PISCES', *Boiling Point* 58: 28 <www.hedon.info/View+issue&itemId=8724>.

Southern Africa Development Committee (SADC) (2010) *SADC Regional Energy Access Strategy and Action Plan*, Eschborn: EUEI-PDF <www.euei-pdf.org/regional-studies/regional-energy-access-strategy-and-action-plan> [accessed 27 June 2013].

Sustainable Energy for All (SE4All) (no date) 'Objectives', Washington, DC: United Nations Foundation <www.sustainableenergyforall.org/objectives> [accessed 2 May 2013].

Sustainable Energy Market Acceleration (SEMA) (no date) [website] <www.sema-project.org/> [accessed 2 May 2013].

Tsephel, S., Takama, T., Lambe, F. and Johnson, F.X. (2009) 'Why perfect stoves are not always chosen: a new approach for understanding stove and fuel choice at the household level', *Boiling Point* 57: 6–8 <www.hedon.info/docs/BP57_LambeEtAl.pdf> [accessed 27 June 2013].

United Nations (UN) (no date) 'Secretary General, at Summit High Level Event, describes energy poverty as obstacle to Millennium Development Goals', New York, NY: UN <www.un.org/News/Press/docs/2010/sgsm13124.doc.htm> [accessed 27 June 2013].

United Nations Development Programme (UNDP) (2012) *Integrating Energy Access and Employment Creation to Accelerate Progress on the Millennium Development Goals in Sub-Saharan Africa*, New York, NY: UNDP <www.undp.org/content/dam/undp/library/Environment%20and%20Energy/Sustainable%20Energy/EnergyAccessAfrica_Web.pdf> [accessed 27 June 2013].

United Nations Environment Programme (UNEP) (2011) 'Solar water heating loan facility in Tunisia (Mediterranean Renewable Energy Programme Finance/PROSOL)' [online] <www.unep.org/climatechange/finance/LoanProgrammes/MEDREP/PROSOLinTunisia/tabid/29559/Default.aspx> [accessed 2 May 2013].

University of Michigan (2002) 'Urbanization and global change', Ann Arbor, MI: University of Michigan <www.globalchange.umich.edu/globalchange2/current/lectures/urban_gc/> [accessed 27 June 2013].

Van der Vleuten, F., Stam, N., and van der Plas, R. (2007) 'Putting solar home system programmes into perspective: what lessons are relevant?', *Energy Policy* 35: 1439–51 <http://dx.doi.org/10.1016/j.enpol.2006.04.001>.

Vermeulen, S. (2001) 'Woodfuel in Africa: crisis or adaptation?', in *Workshop Proceedings: Fuelwood – Crisis or Balance?*, Marstrand, Sweden, 6–8 June.

Vivid Economics (2009) *Advance Market Commitments for Low-Carbon Development: An Economic Assessment*, London: DFID <http://r4d.dfid.gov.uk/PDF/Outputs/EcoDev_Misc/60743-Vivid_Econ_AMCs.pdf> [accessed 22 July 2013].

Wallmo, K. and Jacobson, S.K (1998) 'A social and environmental evaluation of fuel-efficient cook-stoves and conservation in Uganda', *Environmental Conservation* 25: 99–108.

Wilson, E., Godfrey Wood, R. and Garside, B. (2012) *Sustainable Energy for All: Linking Poor Communities to Modern Energy Services*, Working Paper 1, London: IIED.

Wilson, E., Zarsky, L., Shaad, B. and Bundock, B. (2008) 'Lights on or trade off? Can base-of-the-pyramid approaches deliver solutions to energy poverty?', in P. Kandachar. and M. Halme (eds), *Sustainability Challenges and Solutions at the Base of the Pyramid: Business, Technology and the Poor*, Sheffield: Greenleaf Publishing.

Winrock International India (2010) *Access to Clean Energy: a Glimpse of Off-Grid Projects in India*, New Delhi: Government of India, UNDP and Swiss Agency for Development and Cooperation <www.undp.org/content/dam/india/docs/access_to_clean_energy.pdf> [accessed 27 June 2013].

World Bank (no date) 'Renewable Energy Project Toolkit for World Bank Task Managers Case Study: Sri Lanka Renewable Energy Program' <http://siteresources.

worldbank.org/EXTRENENERGYTK/Resources/5138246-1238175210723/
Sri0Lanka0RE0Program0village0hydro0.pdf> [accessed 2 May 2013].

World Bank (2008) *Reforming Energy Prices Subsidies and Reinforcing Social
Protection: Some Design Issues*, Report 43173-MNA, Washington, DC: World
Bank <http://documents.worldbank.org/curated/en/2008/07/9685085/
morocco-reforming-energy-price-subsidies-reinforcing-social-protection-
some-design-issues-reformer-les-subventions-au-prix-de-lenergie-renforcer-
la-protection-sociale-quelques-questions-de-conception#> [accessed 22
July 2013].

World Bank (2010) *Carbon Finance for Sustainable Development: 2010 Annual
Report*, Washington, DC: World Bank <http://siteresources.worldbank.org/
INTCARBONFINANCE/Resources/64897_World_Bank_web_lower_Res..
pdf> [accessed 1 May 2013].

World Bank (2012) 'Mobile phone access reaches three quarters of planet's
population' [press release], July 17 <www.worldbank.org/en/news/press-
release/2012/07/17/mobile-phone-access-reaches-three-quarters-planets-
population> [accessed 27 June 2013].

World Energy Council (WEC) (2010) *2010 Survey of Energy Resources*, London:
WEC <www.worldenergy.org/publications/3040.asp> [accessed 27 June
2013].

Yadoo, A. (2012) *Delivery Models for Decentralised Rural Electrification: Case
Studies in Nepal, Peru and Kenya*, London: IIED.

Yadoo, A. and Cruickshank, H. (2010) 'The value of cooperatives in rural
electrification', *Energy Policy* 38: 2941–7 <http://dx.doi.org/10.1016/j.
enpol.2010.01.031>.

Yoohoon, A. (2010) *Low Carbon Development Path for Asia and the Pacific:
Challenges and Opportunities to the Energy Sector*, ESCAP Energy Resources
Development Series No. 41, ST/ESCAP/2589, Thailand: UN <www.unescap.
org/esd/publications/energy/Series/2010/energy-publication-december-
2010.pdf> [accessed 1 May 2013].

Zerriffi, H. and Wilson E. (2010) 'Leapfrogging over development? Promoting
rural renewables for climate change mitigation', *Energy Policy* 38: 1689–
1700 <http://dx.doi.org/10.1016/j.enpol.2009.11.026>.

www.ingramcontent.com/pod-product-compliance
Lightning Source LLC
Chambersburg PA
CBHW052012030426
42334CB00029BA/3184